# Lecture Notes in Networks and Systems

## Volume 51

**Series editor**

Janusz Kacprzyk, Polish Academy of Sciences, Warsaw, Poland
e-mail: kacprzyk@ibspan.waw.pl

The series "Lecture Notes in Networks and Systems" publishes the latest developments in Networks and Systems—quickly, informally and with high quality. Original research reported in proceedings and post-proceedings represents the core of LNNS.

Volumes published in LNNS embrace all aspects and subfields of, as well as new challenges in, Networks and Systems.

The series contains proceedings and edited volumes in systems and networks, spanning the areas of Cyber-Physical Systems, Autonomous Systems, Sensor Networks, Control Systems, Energy Systems, Automotive Systems, Biological Systems, Vehicular Networking and Connected Vehicles, Aerospace Systems, Automation, Manufacturing, Smart Grids, Nonlinear Systems, Power Systems, Robotics, Social Systems, Economic Systems and other. Of particular value to both the contributors and the readership are the short publication timeframe and the world-wide distribution and exposure which enable both a wide and rapid dissemination of research output.

The series covers the theory, applications, and perspectives on the state of the art and future developments relevant to systems and networks, decision making, control, complex processes and related areas, as embedded in the fields of interdisciplinary and applied sciences, engineering, computer science, physics, economics, social, and life sciences, as well as the paradigms and methodologies behind them.

More information about this series at http://www.springer.com/series/15179

Elżbieta Macioszek · Grzegorz Sierpiński
Editors

# Directions of Development of Transport Networks and Traffic Engineering

15th Scientific and Technical Conference
"Transport Systems. Theory and Practice 2018",
Katowice, Poland, September 17–19, 2018,
Selected Papers

 Springer

*Editors*
Elżbieta Macioszek
Faculty of Transport
Silesian University of Technology
Katowice, Poland

Grzegorz Sierpiński
Faculty of Transport
Silesian University of Technology
Katowice, Poland

ISSN 2367-3370    ISSN 2367-3389 (electronic)
Lecture Notes in Networks and Systems
ISBN 978-3-319-98614-2    ISBN 978-3-319-98615-9 (eBook)
https://doi.org/10.1007/978-3-319-98615-9

Library of Congress Control Number: 2018950540

This Springer imprint is published by the registered company Springer Nature Switzerland AG
The registered company address is: Gewerbestrasse 11, 6330 Cham, Switzerland

# Preface

Transport is one of the most important factors determining the economic development of each country. The implementation of each transport process, both passengers and goods transport, is carried out using transport networks and complementing them with various types of infrastructure. Contemporary technical and technological development, dynamics of changes in the modern world, complexity of road infrastructure as well as the ever-growing mobility of contemporary society provide favourable conditions for designing innovative transport-related solutions. Consequently, activities in the field of transport networks and systems require new decisions which are continuously made with regard to planning, organization, management and traffic control. These decisions entail the necessity of seeking increasingly efficient solutions, undertaking numerous multifaceted activities aimed at meeting the current needs of the society as well as, above all, they determine the directions of development of transport networks and traffic engineering.

The book, entitled *Directions of development of transport networks and traffic engineering*, provides an excellent opportunity to become familiar with the latest trends of research in the field of transport networks and systems, achievements and perspectives for development of these networks, both in Poland and in other countries of the world. In addition, the subjects of the presented works also allow to become familiar with the challenges and innovative directions of research and development in the field of traffic engineering. The book has been divided into four parts. They are as follows:

- Part 1. Human Factors in Traffic Engineering,
- Part 2. Directions of Improvement in Transport Systems,
- Part 3. Traffic Engineering as Support for Development of Transport Networks and Systems,
- Part 4. Modelling Tools in Service of Development of Traffic Engineering.

The publication contains selected papers submitted to and presented at the 15th Scientific and Technical Conference "Transport Systems. Theory and Practice", organized by the Department of Transport Systems and Traffic Engineering at the

Faculty of Transport of the Silesian University of Technology (Katowice, Poland). The topics addressed in the book include the current problems of transport networks and systems, among other subjects discussed. With reference to numerous practical examples, various novel solutions determining the directions of development of transport networks and traffic engineering have been proposed in the publication. They are considered to exert significant influence on increasing the functional efficiency of transport networks and systems, and their priorities include well-being and health of people, traffic safety, sustainable development of transport systems and protection of natural environment. Accordingly, a part of the book also relates to the problems associated with human factors in traffic engineering. Moreover papers also deals with modelling and simulation tools in service of development of traffic engineering were also included in the presented volume.

We would like to use this occasion to express our gratitude to the authors for the papers they have submitted and their substantial contribution to the display directions of development and the multiple challenges facing transport networks and traffic engineering in the contemporary world as well as for rendering the results of their research and sharing the scientific work. We would also like to thank the reviewers for their insightful remarks and suggestions which have ensured the high quality of the publication. Readers interested in the latest achievements and directions of traffic engineering development will find in the volume a comprehensive material presenting the results of scientific research, diverse insights and comments as well as new approaches and problem solutions. With the foregoing in mind, we wish all readers a successful reading.

September 2018                                                                    Elżbieta Macioszek
                                                                                Grzegorz Sierpiński

# Organization

15th Scientific and Technical Conference "Transport Systems. Theoryand Practice" (TSTP2018) is organized by the Department of Transport Systems and Traffic Engineering, Faculty of Transport, Silesian University of Technology (Poland).

## Organizing Committee

### Organizing Chair

Grzegorz Sierpiński                Silesian University of Technology, Poland

### Members

Renata Żochowska
Grzegorz Karoń
Krzysztof Krawiec
Aleksander Sobota
Marcin Staniek
Ireneusz Celiński

Barbara Borówka
Kazimierz Dąbała
Marcin J. Kłos
Damian Lach
Piotr Soczówka

## The Conference Took Place Under the Honorary Patronage

Marshal of the Silesian Voivodeship

# Scientific Committee

## Chairman

| | |
|---|---|
| Stanisław Krawiec | Silesian University of Technology, Poland |
| Rahmi Akçelik | SIDRA SOLUTIONS, Australia |
| Tomasz Ambroziak | Warsaw University of Technology, Poland |
| Henryk Bałuch | The Railway Institute, Poland |
| Roman Bańczyk | Voivodeship Centre of Road Traffic in Katowice, Poland |
| Werner Brilon | Ruhr-University Bochum, Germany |
| Margarida Coelho | University of Aveiro, Portugal |
| Boris Davydov | Far Eastern State Transport University, Khabarovsk, Russia |
| Mehmet Dikmen | Baskent University, Turkey |
| Domokos Esztergár-Kiss | Budapest University of Technology and Economics, Hungary |
| József Gál | University of Szeged, Hungary |
| Andrzej S. Grzelakowski | Gdynia Maritime University, Poland |
| Mehmet Serdar Güzel | Ankara University, Turkey |
| József Hansel | AGH University of Science and Technology Cracow, Poland |
| Libor Ižvolt | University of Žilina, Slovakia |
| Marianna Jacyna | Warsaw University of Technology, Poland |
| Nan Kang | Tokyo University of Science, Japan |
| Jan Kempa | University of Technology and Life Sciences in Bydgoszcz, Poland |
| Michael Koniordos | Pireaus University of Applied Sciences, Greece |
| Bogusław Łazarz | Silesian University of Technology, Poland |
| Zbigniew Łukasik | Kazimierz Pulaski University of Technology and Humanities in Radom, Poland |
| Michal Maciejewski | Technical University Berlin, Germany |
| Elżbieta Macioszek | Silesian University of Technology, Poland |
| Ján Mandula | Technical University of Košice, Slovakia |
| Sylwester Markusik | Silesian University of Technology, Poland |
| Antonio Masegosa | IKERBASQUE Research Fellow at University of Deusto Bilbao, Spain |
| Agnieszka Merkisz-Guranowska | Poznań University of Technology, Poland |
| Anna Mężyk | Kazimierz Pulaski University of Technology and Humanities in Radom, Poland |
| Maria Michałowska | University of Economics in Katowice, Poland |
| Leszek Mindur | International School of Logistic and Transport in Wrocław, Poland |

# Referees

Rahmi Akçelik
Marek Bauer
Przemysław Borkowski
Werner Brilon
Margarida Coelho
Piotr Czech
Domokos Esztergár-Kiss
Michal Fabian
Barbara Galińska
Róbert Grega
Mehmet Serdar Güzel
Katarzyna Hebel
Nan Kang
Peter Kaššay
Jozef Kuĺka
Michał Maciejewski
Elżbieta Macioszek
Krzysztof Małecki
Martin Mantič

Silvia Medvecká-Beňová
Katarzyna Nosal Hoy
Romanika Okraszewska
Asier Perallos
Hrvoje Pilko
Antonio Pratelli
Michal Puškár
Piotr Rosik
Alžbeta Sapietová
Grzegorz Sierpiński
Marcin Staniek
Dariusz Tłoczyński
Andrzej Więckowski
David Williams
Grzegorz Wojnar
Adam Wolski
Ninoslav Zuber

# Contents

# Human Factors in Traffic Engineering

# Changes in the Level of Satisfaction and Passenger Preferences in Sustainable Public Transport

Zofia Bryniarska[⊠]

Faculty of Civil Engineering, Cracow University of Technology,
Kraków, Poland
z_bryn@pk.edu.pl

**Abstract.** Passenger satisfaction and preference surveys allow for assessing the quality of public transport operation from the point of view of passengers. One of the basic principles that should be followed by organizers of public transport is to achieve sustainable transport in urban areas by offering high-quality services and encouraging as many residents as possible to use this form of mobility. Technical, technological, innovative and quantitative changes introduced by organizers and transport operators are reflected in passengers' perception of public transport services in urban areas. The aim of the article is to show that passengers begin to appreciate transport attributes such as the frequency and the duration of travels more than punctuality and the ability to travel without transferring at interchanges. In addition, the importance of direct interpretation of the quality gap of selected quality attributes by organizers and mass transport operators has been demonstrated.

**Keywords:** Urban public transport · Urban area · Passengers' satisfaction
Passengers' preferences · Transport service quality · CSI · HCSI

## 1 Introduction

Public transport is the best solution in terms of citizens' mobility in urban areas [1, 2]. On the one hand, transport needs are on the increase, but on the other hand, travelers and other transport stakeholders have growing demands. Passengers value their time more and more, and expect that travelling will take as short as possible or that during travel they will be able to work or entertain themselves. But they also have an increasing environmental awareness. The White Paper of 2011 [3] emphasized the vision of a competitive and sustainable transport system and the need to promote integrated mobility while achieving the 60% emission reduction target.

The task of meeting the growing demands of passengers, transport managers and environmental protection is becoming a challenge for today and for the coming years. For individuals, car travel is generally more comfortable, flexible and faster for supporting their busy lifestyles [4]. But public transport in cities has to and can effectively compete with individual transport [5].

© Springer Nature Switzerland AG 2019
E. Macioszek and G. Sierpiński (Eds.): Directions of Development of Transport
Networks and Traffic Engineering, LNNS 51, pp. 3–16, 2019.
https://doi.org/10.1007/978-3-319-98615-9_1

The adoption of a customer-oriented (passenger-oriented) perspective [6], where it is the passenger who knows best what she or he needs, what quality attributes are important to her/him or what is missing for her/him from the offer, have become the key to increasing the attractiveness of public transport. According to this idea, the Transit Capacity and Quality of Service Manual [7] considers the customers' point of view to be the fundamental perspective for service assessment.

The article will present the results of marketing research carried out in 2017 on a sample of over 1200 respondents. The main objective of the study was to confirm or reject the passengers' claim that punctuality is the most important quality attribute and to express whether they appreciate the frequency and the duration of travels more than punctuality and the ability to travel without transferring at interchanges. In the conditions of uncertainty of events and the congestion of city streets, this aspect is exceptionally difficult to meet. The respondents were also asked to express their opinion about satisfaction and preferences for eleven selected quality attributes. Based on these opinions, a quality gap was identified and the customer satisfaction index (CSI) and the heterogeneous customer satisfaction index (HCSI) were calculated.

It is worth emphasizing that passenger satisfaction and preference surveys help adjust the way public transport functions to current passenger needs and can contribute to increasing its role in urban areas and improving the modal split distribution.

The article is composed in six sections. In the beginning the brief literature review of some works concerning the analysis of customer satisfaction and preferences is given. Then the methodology of passenger satisfaction surveys is presented and way of valuating overall satisfaction with transit service quality and two customer satisfaction indexes (CSI and HCSI) are discussed. The fourth section contain the description of survey conducted in Krakow and main characteristics of the sample and the analysis of passengers' satisfaction and preferences rates and evaluation of overall satisfaction with service quality of public transport. In the next section the presentation of analysis and discussion of their results is elaborated. Finally the author propose a brief sum up and discussion of obtained results.

## 2    Literature Review

In the literature on the subject, many studies can be found regarding the measurement and analysis of the quality of public transport services from the point of view of its users. The best known and most widely applied technique is the ServQual method [8], where the concept of customer satisfaction as a function of customer expectations (what customers expect from the service) and perceptions (what customers receive) has been introduced. This method also provides the index in form of difference between satisfaction/perception and expectation/preferences for quality attributes. Later, certain variations of this method have been introduced called ServPerf method [9] or Normed Quality method [10].

Other authors developed and explored the approach based on factor analysis, ordered logit modelling or multinomial logit model. This way, they [11] analyzed the variability of the user's behavior and their level of satisfaction with the use of diverse transit systems. The latent variables models help capture unobserved variables in describing perception and attitudes [12].

Structural equation models are formulated in order to explore relationships between various aspects and attributes of transport quality and global customer satisfaction [13–15]. Through this type of model, the strength of the relationships can be quantified and compared with one another in terms of both direct and indirect effects.

The next approach is based on service quality visible in the user perceptions expressed in terms of choice. The interviewed users make a choice from among some alternative services characterized by certain service quality attributes varying on more levels. The solution was proposed in form of logit models [16, 17], nested logit models [18] or probit models [19].

The other methodology that was based on the use of both passenger perceptions and transit agency performance measures involving the main aspects that characterize a transit service is also developed [20]. The combination of these two types of service quality measure is understand as a useful and reliable measurement tool of transit performance that could be the most suitable solution for providing a real understanding of the phenomena analyzed.

## 3 Methodology

### 3.1 Passenger Satisfaction Surveys

Customer satisfaction represents a measure of company performance according to customer needs [21]. Therefore, the measure of customer satisfaction provides a service quality measure. In the customer survey, passengers are asked and can express their judgment regarding selected aspects of the transport service ad hoc. Usually, various aspects that characterize the service, e.g. reliability, comfort, spatial accessibility and time availability, information, cleanness in vehicles, frequency, safety and security, are examined and analyzed. Passengers express their opinions in Likert scale on five or seven levels of agreement/disagreement. Sometimes they can use school-scale (0–5 or 1–6, possibly a percentage scale 0–100%).

Passengers are also questioned about the preferences/importance of particular aspects of service provision. In this way, they can express their approach to the importance and hierarchy of individual traits.

The analysis of the difference between the assessment of passenger satisfaction and preferences determines the scope and value of the discrepancy between transport service provision and the requirements of passengers as to what the service should look like. Therefore, it is important to know the needs of passengers, their preferences and satisfaction with solutions in the field of public transport operation and to eliminate the identified quality gap.

## 3.2  Overall Satisfaction with Transit Service Quality and Customer Satisfaction Index

The overall transit service quality level can be expressed by passengers during the survey. In this case, a 10-item or a percentage scale is used, which is often easier for respondents to understand, accept and then express their opinion. Questions are asked at the beginning of the survey process, before asking about detailed aspects of the quality attributes of the transport service. Some researchers believe that it is worth repeating the question at the end, that is, after having received individual quality assessment from the respondents. The overall transit service quality assessed at that time is usually higher.

Customer Satisfaction Index (CSI) [21] represents a measure of service quality on the basis of the user/consumer perceptions of service aspects expressed in terms of importance rates $I_k$, compared with user/consumer expectations expressed in terms of satisfaction rates $S_k$. It is calculated in accordance with the formula:

$$CSI = \sum_{k=1}^{N} \left[ \bar{S}_k \cdot W_k \right] \tag{1}$$

in which $\bar{S}_k$ is the mean of the satisfaction rates expressed by users on the service quality $k$ attribute and $W_k$ is a weight of the $k$ attribute, calculated on the basis of the importance rates expressed by users and is calculated using the following formula:

$$W_k \frac{\bar{I}_k}{\sum_{k=1}^{N} \bar{I}_k} \tag{2}$$

CSI does not take into account the heterogeneities among user judgments. The heterogeneous customer satisfaction index HCSI formula overcomes this disadvantage [22]. The HCSI index is calculated using the formula:

$$HCSI = \sum_{k=1}^{N} \left[ S_k^c \cdot W_k^c \right] \tag{3}$$

in which $S_k^c$ is the mean of the satisfaction rates expressed by users on the $k$ attribute corrected according to the deviation of the rates from the average value and is calculated using the following formula:

$$S_k^c = \bar{S}_k \frac{\frac{\bar{S}_k}{\text{var}(S_k)}}{\sum_{k=1}^{N} \frac{\bar{S}_k}{\text{var}(S_k)}} N \tag{4}$$

and $W_k^c$ is the weight of the $k$ attribute, calculated on the basis of the importance rates expressed by users, corrected according to the dispersion of the rates from the average value and is calculated using the following formula:

$$W_k^c = \frac{\frac{\bar{I}_k}{\mathrm{var}(I_k)}}{\sum_{k=1}^{N} \frac{\bar{I}_k}{\mathrm{var}(I_k)}} \tag{5}$$

# 4 The Results of Survey in Krakow

## 4.1 Characteristics of Survey

The survey was carried out by students of the second-cycle studies in the Transport major at the Cracow University of Technology in October and November 2017. Most surveys were conducted face to face with passengers at the largest communication stops (848 questionnaires - 68.8%), and the remaining ones in the form of an online survey (384 questionnaires - 31.2%).

There were three groups of questions in the specially designed questionnaire. In the first group, there was a question about the overall transit service quality of public transport in Krakow and three questions in which the respondent had to choose what quality attribute of public transport was more important to her/him: (a) punctuality or frequency, (b) travel time or opportunity to travel without having to change, (c) travel duration or conditions of traveling in the vehicle (in particular, overcrowding of the vehicle). The assessment of overall transit service quality consisted in determining the subjective feelings of the respondent on a scale from 0 to 100%, where 0% meant total dissatisfaction, and 100% the highest level of satisfaction.

The second group of questions included the assessment of satisfaction and preferences from selected quality attributes of public transport. The eleven attributes that were selected for assessment were used in surveys carried out in previous years [23]. They included the following attributes: (a) frequency of running, (b) punctuality, (c) direct connections (without change), (d) convenience of transfer, (e) information at stops and in vehicles, (f) conditions of travel in vehicles, comfort of driving, (g) conditions of waiting at stops, (h) safety at the bus stop and in the vehicle, (i) running rhythmicity, (j) reliability of the planned journey, (k) speed of travel/duration of the journey. The respondents were asked to make an assessment by expressing their satisfaction with the actual provision of mass transport services and then their preferences (importance) of a given feature for the respondent. Satisfaction and preference assessments were to be expressed on a scale from 0 to 5, where 1 means a very low degree, and 5 a very high degree of satisfaction or preference.

The third group of questions was aimed to define the socio-demographic characteristics of the respondent and their way of using public transport. The socio-demographic situation was identified based on gender, age (12–18, 19–26, 27–44, 45–60/65, over 60/65) and professional status (school student, higher education student, mobile worker, immobile worker, senior citizen or pensioner). The way of using public transport was determined on the basis of: (a) frequency of using public transport (daily, several times a week, several times a month, occasionally), (b) defining the average time that the passenger travels by means of public transport (up to 20 min, 20–40 min, 40–60 min, over 60 min), (c) tickets used during the travel (single ticket, short-term or season tickets or free rides). Season tickets or free journeys were treated jointly, because they give passengers a similar feeling of freedom to choose more vehicles and transfer, without having to pay additional fees.

## 4.2 Case Study

The study area is a city of Krakow, located in the south of Poland. It is the second largest city in terms of population, an important economic, cultural and tourist center, with several universities and academies. The urban area has approximately 760 000 inhabitants but during the academic year, there are about 200,000 additional residents - students.

The public transport system consists of 23 tram lines, 79 inner-city bus lines and 66 agglomeration lines that serve the area of communes adjacent to the city. Tram lines have a length of 4.9 to 21.7 km, during the survey process they were running in the peak hours with the frequency of 5 min (2 lines), 10 min (10 lines) and 20 min (11 lines). The inner city bus lines have a length from 3.0 to 22.5 km. The frequency of running is from 10–30 min. The agglomeration lines are between 5.2 and 38.4 km long. Most of them run every 30, 60 or 90 min.

On a typical working day, the number of passengers exceeds 1.2 million, of which about 0.68 million are passengers of tram transport. The average length of journey by tram is 3.42 km, by bus within the city limits 3.87 km, and by agglomeration lines 7.17 km. The average travel duration is 11.6, 11.9 and 16.2 min, respectively [24].

## 4.3 Socio-Demographic Characteristics of Survey Respondents

The survey was addressed to users of urban tram and bus transport in Krakow. The general characteristics of the respondents are presented in Table 1.

The survey participants were mainly women (58.2%), aged 19–26 (51.1%), most often studying at one of universities (49.4%). People at this age often do not have their own means of transport, they are mobile, during the typical working day they travel several times between places where their classes are held, their place of residence and recreation. It is worth noting that many of the respondents were also mobile and immobile workers (32.6%), and pensioners (8.9%).

Respondents most often declared that they use public transport every day (65.1%) or several times a week (18.9%). The duration of their journeys is evenly distributed (approximately 30%) for periods up to 20 min, 20–40 min and 40–60 min. Only 7.8% are journeys longer than 60 min. The most frequently used tickets are season tickets or free travel (58.2%).

**Table 1.** General characteristics of respondents ($n = 1232$).

| Characteristics | Example | Volume | % |
|---|---|---|---|
| Gender | Female | 717 | 58.2 |
| | Male | 515 | 41.8 |
| Age | 12–18 years | 109 | 8.8 |
| | 19–26 years | 629 | 51.1 |
| | 27–44 years | 192 | 15.6 |
| | 45–60/65 years | 213 | 17.3 |
| | Over 60/65 years | 89 | 7.2 |
| Professional status | Pupil | 111 | 9.0 |
| | Student | 609 | 49.4 |
| | Mobile worker | 122 | 9.9 |
| | Immobile worker | 280 | 22.7 |
| | Senior citizen/Pensioner | 110 | 8.9 |
| Frequency of using public transport | Every day | 802 | 65.1 |
| | A few times a week | 233 | 18.9 |
| | A few times a month | 91 | 7.4 |
| | Occasionally | 106 | 8.6 |
| Duration of journey | Up to 20 min | 360 | 29.2 |
| | 20–40 min | 404 | 32.8 |
| | 40–60 min | 372 | 30.2 |
| | Over 60 min | 96 | 7.8 |
| Type of ticket | Single | 202 | 16.4 |
| | Short-term | 313 | 25.4 |
| | Season ticket or free travel | 717 | 58.2 |

## 4.4 Validation of Selected Quality Attributes

The results of surveys carried out since the 1990s in Polish cities show that punctuality was the most important quality attribute for passengers [25]. Moreover, passengers highly appreciated the possibility of using journeys without the necessity of transferring between lines or modes of transport, even at the expense of longer waiting times at the bus/tram stop for the right vehicle. In recent years, a lot has changed. The basic bus and tram lines operate at a frequency of less than 10 min, an even better situation takes place on the basic communication routes on which subsequent vehicles run every 1–2 min. Moreover, other important advantages include dynamic passenger information at tram stops or mobile applications that show on-line information about the expected time of arrival of the vehicle, low floor in vehicles and the construction of stop platforms to facilitate boarding and alighting.

During the survey, passengers were also directly asked these questions. The results of the respondents' declarations are presented in Table 2. Furthermore, the distribution of responses depending on the frequency of using public transport, age, occupational status and average journey time is demonstrated here.

**Table 2.** Validation of selected quality attributes ($n = 1232$).

| Criteria | Questions | | | | | |
| --- | --- | --- | --- | --- | --- | --- |
| | Frequency is more important than punctuality | | Duration of the trip is more important than possibility to travel without interchanging | | Duration of the trip is more important than comfort during travel | |
| | Yes | No | Yes | No | Yes | No |
| All respondents | 59.8 | 40.2 | 56.8 | 43.2 | 69.4 | 30.6 |
| Frequency of using public transport | | | | | | |
| Every day | 40.7 | 24.4 | 40.5 | 24.6 | 47.0 | 18.1 |
| A few times a week | 10.5 | 8.4 | 8.7 | 10.2 | 12.2 | 6.7 |
| A few times a month | 4.1 | 3.3 | 3.6 | 3.8 | 4.5 | 2.8 |
| Occasionally | 4.6 | 4.0 | 4.1 | 4.5 | 5.7 | 2.9 |
| Professional status | | | | | | |
| Pupil | 5.3 | 3.6 | 5.8 | 3.1 | 6.6 | 2.3 |
| Student | 32.6 | 18.4 | 31.3 | 19.7 | 37.7 | 13.3 |
| Mobile worker | 8.8 | 6.8 | 8.9 | 6.7 | 11.3 | 4.3 |
| Immobile worker | 9.3 | 8.0 | 8.3 | 9.0 | 10.9 | 6.4 |
| Senior citizen/Pensioner | 3.9 | 3.3 | 2.5 | 4.7 | 2.9 | 4.3 |
| Age | | | | | | |
| 12–18 years | 5.4 | 3.7 | 5.8 | 3.2 | 6.7 | 2.4 |
| 19–26 years | 31.7 | 17.8 | 30.5 | 18.9 | 36.8 | 12.7 |
| 27–44 years | 6.1 | 3.8 | 5.4 | 4.5 | 7.2 | 2.7 |
| 45–60/65 years | 11.9 | 10.8 | 12.3 | 10.4 | 14.9 | 7.8 |
| Over 60/65 years | 4.8 | 4.1 | 2.8 | 6.2 | 3.8 | 5.1 |
| Duration of journey | | | | | | |
| Up to 20 min | 17.6 | 11.6 | 15.6 | 13.6 | 19.7 | 9.5 |
| 20–40 min | 19.6 | 13.2 | 19.1 | 13.7 | 22.6 | 10.1 |
| 40–60 min | 18.4 | 11.8 | 17.7 | 12.5 | 21.8 | 8.4 |
| Over 60 min | 4.2 | 3.6 | 4.5 | 3.3 | 5.3 | 2.5 |

Respondents strongly indicated that the frequency of running (59.8%) is more important to them than punctuality. Moreover, such a position is always predominant in each group (groups distinguished by the frequency of using public transport, age, professional status or duration of travel).

The situation is slightly different when respondents are asked about the importance of the duration of the journey relative to the possibility of traveling without interchanging. Most of the respondents emphasize that travel duration is more important to them (56.8%) than the possibility of traveling without interchanging. However, in the subdivision groups: those traveling several times a month or occasionally, in the professional status group of non-mobile workers and pensioners and people aged over 60/65 declare that they prefer traveling without interchanging. It is clear that older people (over 60/65 years of age) are no longer physically active and choose transport behaviors that are more comfortable to them even at the expense of travel time.

The vast majority of respondents (69.4%) attribute greater value to the shorter duration than the convenience and comfort of travel. In this case, only people aged 60/65+ and retired people declared differently.

## 4.5    Assessment of Passengers' Satisfaction and Preferences

The respondents were asked to assess the satisfaction and preferences of eleven selected quality attributes of public transport. The summary of satisfaction and preference assessment is presented in Table 3 in the form of average values for each quality feature indicating the average value, standard deviation being a measure of the variability of this assessment, and the confidence interval (for a confidence coefficient of 0.95).

**Table 3.** Satisfaction and preference statistics ($n = 1232$).

| Service Attribute | No | Satisfaction | | | | Preferences | | | |
|---|---|---|---|---|---|---|---|---|---|
| | | Mean | St.dev | Conf Inf | | Mean | St.dev | Conf Inf | |
| Frequency of running | 1 | 3.45 | 0.82 | 3.40 | 3.49 | 4.46 | 0.73 | 4.42 | 4.50 |
| Punctuality of running | 2 | 3.15 | 0.95 | 3.09 | 3.20 | 4.44 | 0.77 | 4.39 | 4.48 |
| Direct connections | 3 | 3.43 | 0.89 | 3.38 | 3.47 | 4.16 | 0.87 | 4.12 | 4.21 |
| Convenience of transfer | 4 | 3.44 | 0.91 | 3.39 | 3.49 | 4.09 | 0.82 | 4.05 | 4.14 |
| Information at stops and in vehicles | 5 | 3.99 | 0.91 | 3.94 | 4.04 | 4.05 | 0.97 | 3.99 | 4.10 |
| Conditions of travel in vehicles, comfort of driving | 6 | 3.37 | 0.94 | 3.32 | 3.42 | 4.09 | 0.84 | 4.04 | 4.13 |
| Conditions of waiting at stops | 7 | 3.32 | 0.93 | 3.27 | 3.38 | 3.99 | 0.96 | 3.94 | 4.04 |
| Safety at bus stop and in the vehicle | 8 | 3.64 | 0.94 | 3.59 | 3.69 | 4.30 | 0.84 | 4.26 | 4.35 |
| Running rhythmicity | 9 | 3.40 | 0.96 | 3.35 | 3.46 | 4.05 | 0.92 | 4.00 | 4.10 |
| Reliability of the planned journey | 10 | 3.45 | 0.99 | 3.40 | 3.51 | 4.61 | 0.68 | 4.58 | 4.65 |
| Speed of travel/duration of the journey | 11 | 3.25 | 0.95 | 3.20 | 3.30 | 4.46 | 0.72 | 4.42 | 4.50 |

Passengers allocated the highest satisfaction score to information at stops and in vehicles (3.99) and to safety at the bus/tram stops and in vehicles (3.64). The lowest satisfaction score was attributed to the punctuality of running (3.15) and the duration of the journey (3.25). Although volatility measures for individual quality attributes do not exceed 0.99 (coefficient of variation is 0.23–0.30), differences in the scores are clearly visible, but it indicates a significant homogeneity of respondent assessments.

The respondent preference assessments (importance of attributes) show that the reliability of a planned trip (4.61), frequency of running (4.46) and speed of travel/duration of the journey (4.46) are most important to passengers, while the waiting conditions at stops (3.99), information at stops and in vehicles (4.05) and running rhythmicity (4.05) are the least important. The score diversification of these attributes is much smaller.

## 4.6    Overall Satisfaction with Service Quality of Public Transport

The respondents were asked to assess their overall satisfaction with service quality of public transport functioning. The scores were to be expressed on a percentage scale from 0 to 100. The results of respondents' answers are recorded in Table 4.

**Table 4.** Descriptive statistics of overall transit service quality.

| Scale of satisfaction with transit service quality 0–100% | Mean | Quartile | | | Standard deviation | Coefficient of variation [%] | Skewness | Kurtosis |
|---|---|---|---|---|---|---|---|---|
| | | $Q_1$ | $Q_2$ | $Q_3$ | | | | |
| 10-item | 65 | 60 | 70 | 80 | 16.9 | 26.0 | −0.76 | 0.66 |

The majority of respondents (28.4%) declared 70% satisfaction with the functioning of public transport. More than 50% of respondents indicated scores from 60% to 80%. The average value of satisfaction with the overall transit service quality is 65%. The distribution of responses shows strong concentration and left-skewness, which means that the most respondents are slightly more satisfied than on average.

The value of the general assessment of the overall transit service quality can also be carried out using CSI or HCSI indexes. Prior to the determination of both indexes, the internal conformity of the assessment of selected quality features was examined. The α-Cronbach coefficient was used [26]. The result alpha = 0.837 confirms that the group of selected attributes is internally consistent. The calculation results of CSI and HCSI indexes are presented in Table 5.

The CSI index reaches 3.44, while the HCSI index is slightly lower and amounts to 3.41. The values of the two indexes are similar, since the respondents' scores are relatively homogeneous and do not exhibit significant heterogeneity.

The previously determined average value of overall satisfaction with service quality of public transport on the basis of respondents' answers on a scale of 0–100% amounted to 65%. On a 5-point scale, it would have to be converted into 3.25. It should also be noted that the respondents answered the question about the overall transit service quality earlier than questions about the level of satisfaction with and preferences as to selected quality attributes. The scores obtained on the basis of CSI and HCSI indexes are higher. As many authors emphasize [22], the comparison of the overall transit service quality of respondents formulated before and after the evaluation of individual quality features differs. The ratings declared after expressing overall transit service quality score in relation to individual quality attributes are higher. In this survey study, the assessment based on CSI and HCSI indices confirms this regularity.

**Table 5.** Calculation of CSI and HCSI.

| Service attribute | No | Importance weight | Weighted score | Corrected importance weight | Corrected satisfaction | Weighted score |
|---|---|---|---|---|---|---|
| Frequency of running | 1 | 0.096 | 0.33 | 0.118 | 4.36 | 0.51 |
| Punctuality of running | 2 | 0.095 | 0.30 | 0.105 | 2.70 | 0.28 |
| Direct connections | 3 | 0.089 | 0.31 | 0.076 | 3.68 | 0.28 |
| Convenience of transfer | 4 | 0.088 | 0.30 | 0.085 | 3.52 | 0.30 |
| Information at stops and in vehicles | 5 | 0.087 | 0.35 | 0.060 | 4.79 | 0.29 |
| Conditions of travel in vehicles, comfort of driving | 6 | 0.087 | 0.29 | 0.081 | 3.18 | 0.26 |
| Conditions of waiting at stops | 7 | 0.085 | 0.28 | 0.061 | 3.16 | 0.19 |
| Safety at bus stop and in vehicle | 8 | 0.092 | 0.34 | 0.086 | 3.74 | 0.32 |
| Running rhythmicity | 9 | 0.087 | 0.30 | 0.067 | 3.08 | 0.21 |
| Reliability of the planned journey | 10 | 0.099 | 0.34 | 0.140 | 2.99 | 0.42 |
| Speed of travel/duration of the journey | 11 | 0.121 | 0.31 | 0.121 | 2.89 | 0.35 |
|  |  | CS Index | 3.44 |  | HCS Index | 3.41 |

## 5 Presentation of Analysis and Discussion

The analysis of the satisfaction and preference assessment scores of selected quality attributes of public transport allows for determining the gap score - the difference between these values. The satisfaction and preference scores and the quality gap scores arranged according to the increasing value of these gap scores are presented in Fig. 1. The specification of the quality gap score is a convenient and tangible way of showing for which quality attributes the assessment of the existing real situation, from the point of view of passengers, departs from their assessment of the importance of this attribute. The positive values of the quality gap score indicate how intensive should be the actions that should be taken by the public transport organizer or operator to improve the quality assessment of public transport performance in passengers' opinions.

The largest quality gap score (1.291) was identified for the punctuality of running and the duration of the journey (1.209). In the case of these two attributes, passengers' expectations are the highest.

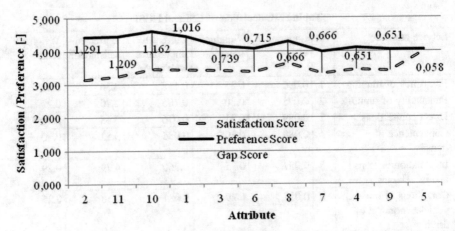

**Fig. 1.** Quality gap for eleven selected characteristics of the quality of public transport

The lowest passenger expectations concern passenger information at stops and in vehicles. Technological and technical innovations thanks to which passengers have access to information about the real time of arrivals of vehicles (currently only at tram transport stops or in mobile applications) were expressed in the respondents' opinions.

## 6  Conclusions

The functioning and development of public transport are currently seen as the basic activities/measures that can improve the quality of life of residents in urban areas. On the one hand, they allow for providing effective methods for moving within the city, and, on the other hand, for reducing energy consumption and transport costs as well as external costs (pollution, noise, land use) [26] for current and future generations.

It is worth emphasizing that in order to encourage residents to use means of public transport and keep those who are currently using them, there is the strong necessity to listen to the opinions and feelings of passengers. The measurement and monitoring of public transport quality is an important issue for organizers and transport operators [27]. They should constantly observe changes in the quality of transport services provided and identify those quality attributes that are unsatisfactory according to passengers. Such a function is fulfilled by the quality gap score set in the survey study for individual quality attributes, which allows for targeting activities/interventions to the right issues. Such an action improves the level of passenger satisfaction score [28].

Additionally, a detailed analysis of the transport quality attributes shows that passengers now declare that they attach greater importance to the frequency than to the punctuality of running in all surveyed groups of respondents. Moreover, the frequency of running is more important than the possibility of traveling without interchanging, yet this declaration rather applies to young people.

The high scores of passenger preferences as to the frequency of running and the reliability of the planned journey identified in the survey should prompt the organizers and transport operators to focus on providing high frequency and preventing situations that reduce the reliability of transport operation. The higher the passenger satisfaction scores as to quality attributes, the more effective the overall system is and the more passengers would travel by means of public transport. In the authors opinion further research should be carried out in learning on the one hand how extension of public transport service accessibility and availability and on the other hand development of new services (light railway service) influence on passengers perception of urban public transport service quality.

# References

1. Dziadek, S.: Systemy Transportowe Ośrodków Zurbanizowanych. PWN, Warszawa (1991)
2. Stjernborg, V., Mattisson, O.: The role of public transport in society - a case study of general policy documents in Sweden. Sustainability 8(11), 1–16 (2016)
3. European Commission: White Paper. Roadmap to a Single European Transport Area - Towards a Competitive and Resource Efficient Transport System. Publications Office of the European Union, Luxembirg (2011)
4. Jakobsson Bergstad, C., Gamble, A., Hagman, O., Polk, M., Garling, T.: Affective-symbolic and instrumental-independence psychological motives mediating effects of socio-demographic variables on daily car use. J. Transp. Geogr. 19(1), 33–38 (2011)
5. Bryniarska, Z.: Competition tools in passenger transport in urban areas. In: Macioszek, E., Sierpiński, G. (eds.) Contemporary Challenges of Transport Systems and Traffic Engineering. LNNS, vol. 2, pp. 3–13. Springer, Cham (2017)
6. Drucker, P.F.: The Practice of Management. Harper Business, San Francisco (2006)
7. Transportation Research Board: Transit Capacity and Quality of Service Manual, 2nd edn. Transportation Research Board of the National Academy of Science, Washington (2004)
8. Parasuraman, A., Zeithaml, V.A., Berry, L.L.: A conceptual model of service quality and its implication for future research. J. Mark. 49, 41–50 (1985)
9. Cronin, J.J., Taylor, S.A.: SERVPERF versus SERVQUAL: reconciling performance-based and perceptions-minus-expectations measurement of service quality. J. Mark. 58(1), 125–131 (1994)
10. Teas, R.K.: Expectations, performance evaluation and consumers' perceptions of quality. J. Mark. 57(4), 18–34 (1993)
11. Tyrinopoulos, Y., Antoniou, C.: Public transit user satisfaction: variability and policy omplications. Transp. Policy 15(4), 260–272 (2008)
12. Nurul-Habib, K.M., Kattan, L., Islaam, T.: Why do the People Use Transit? A Model for Explanation of Personal Attitude Towards Transit Service Quality. https://trid.trb.org/view/880540
13. Joewono, T.B., Kubota, H.: User perception of private paratransit operation in Indonesia. J. Public Transp. 10(4), 99–118 (2007)
14. Stuart, K.R., Mednick, M., Bockman, J.: Structural equation model of customer satisfaction for the New York City subway system. Transp. Res. Rec. 1735, 133–137 (2000)
15. Eboli, L., Mazzulla, G.: Service quality attributes affecting customer satisfaction for bustransit. J. Public Transp. 10(3), 21–34 (2007)
16. Hensher, D.A.: Service Quality as a Package: What Does it Mean to Heterogeneous Consumers. https://trid.trb.org/view/698964

17. Chica-Olmoa, J., Gachs-Sánchezb, H., Lizarragac, C.: Route Effect on the Perception of Public Transport Services Quality. http://ocs.editorial.upv.es/index.php/CIT/CIT2016/paper/viewFile/3515/1414
18. Hensher, D.A., Stopper, P., Bullock, P.: Service quality-developing a service quality index in the provision of commercial bus contracts. Transp. Res. Part A 37, 499–517 (2003)
19. Dell'Olio, L., Ibeas, L., Cecín, P.: Modelling user perception of bus transit quality. Transp. Policy 17, 388–397 (2010)
20. Eboli, L., Mazzulla, G.: A methodology for evaluating transit service quality based on subjective and objective measures from the passenger's point of view. Transp. Policy 18, 172–181 (2011)
21. Hill, N., Brierley, G., MacDougall, R.: How to Measure Customer Satisfaction. Gower Publishing, Hampshire (2003)
22. Eboli, L., Mazzulla, G.: A new customer satisfaction index for evaluating transit service quality. J. Public Transp. 12(3), 21–37 (2009)
23. Starowicz, W.: Kształtowanie Jakości Usług Przewozowych w Miejskim Transporcie Zbiorowym. Uniwersytet Szczeciński, Szczecin (2001)
24. Zintegrowany System Taryfowo-Biletowy w Obszarze Aglomeracji Krakowskiej. Etap I - Badania Napełnień Pasażerskich w Pojazdach Komunikacji Miejskiej w Krakowie na Liniach Miejskich. Raport International Management Services Sp. z o.o. dla Zarządu Infrastruktury Komunalnej i Transportu, Kraków (2014)
25. Ciastoń, A., Sapoń, G.: Hierarchizacja Preferencji i Ocen Cech Jakości w Transporcie Zbiorowym na Podstawie Współczynnika W-Kendala. Transp. Miejski i Regionalny 2, 36–39 (2009)
26. Cronbach, L.J.: Coefficient alpha and the internal structure of tests. Psychometrika 16, 297–334 (1951)
27. Korzhenevych, A., Dehnen, N., Bröcker, J., Holtkamp, M., Meier, H., Gibson, G., Varma, A., Cox, V.: Update of the Handbook on External Costs of Transport. Final Report for the European Commission DG Move, United Kingdom (2014)
28. de Oña, R., Eboli, L., Mazzulla, G.: Monitoring changes in transit service quality over time. Procedia – Soc. Behav. Sci. 111, 974–983 (2014)

# Polish Systems of Car-Sharing - The Overview of Business to Customer Service Market

Katarzyna Turoń[✉] and Piotr Czech

Faculty of Transport, Silesian University of Technology, Katowice, Poland
{katarzyna.turon, piotr.czech}@polsl.pl

**Abstract.** This thesis was conducted on one of the sustainable solution in mobility-shared, short-term service of using cars- car-sharing. In view of the growing popularity of car-sharing services in Polish towns and wide variety of functioning systems, the article presents current state of car-sharing systems in Poland. Authors present the review of Polish market. The analysis was performed on entities providing car-sharing recommending their services and their variety in depending upon registration in system, data verification, type of system and method of rental, localization, used fleet vehicles and cost of services.

**Keywords:** Car-sharing systems · Car-sharing systems in Poland
Organized car-sharing · Car-sharing maintenance

## 1 Introduction

Promoting the idea of sustainable urban mobility is related to constant search of opportunities which can persuade users of individual vehicles to change of their traffic behavior in line with leaving the assumptions of cities aimed at car transport [1, 2]. This type of actions are supposed to create new mobility culture (i.a. [3–5]) and contribute to smooth the traffic in cities, reduce the negative impact of vehicles on the environment and to improve the situation of pedestrians and cyclists on the roads (i.a. [6–10]). In line with these trends, transport managers strive to gradually accustom road users to changing modes of transport by using, at various stages of travel, respectively, spare parts for individual vehicles [11]. Thanks to this kind of approach new solutions to last-mile logistic are made in respect to carriage of passengers [12, 13]. One of the solutions may be the assistance of urban transportation system to which belongs all possibilities which are offered by shared mobility solutions [14]. In the case of transport, these possibilities may relate to the sharing of a vehicle (own or rented from a particular operator), travel sharing (ride-sourcing) or sharing their own vehicle on specific travel routes (car-pooling) [15, 16]. Among these three forms of shared mobility, the urban car-sharing services are an option that has been developing strongly in recent years [17].

Following the practice of improving and diversifying mobility in foreign practice, also in Poland, since 2016, short-term rental vehicles in cities have been functioning. The gradual appearance of new operators on the Polish market introduced new forms of services offered. Initially available only in Warsaw and Krakow, services began to reach other Polish cities. Their development also introduced new functionalities related

© Springer Nature Switzerland AG 2019
E. Macioszek and G. Sierpiński (Eds.): Directions of Development of Transport Networks and Traffic Engineering, LNNS 51, pp. 17–26, 2019.
https://doi.org/10.1007/978-3-319-98615-9_2

to technical issues of vehicle rental, forms of vehicle return or started the appearance of electric car fleets. Due to differences in functioning, the authors analyzed the current state of car-sharing systems in Poland. The work focuses on the analysis of the method of registration in the system and verification of user data, type of systems, methods of vehicle rental, size of vehicle fleet and service costs. In addition, the type of vehicles used has been analyzed, taking into account their segments and the models used. The analysis focused on organized operators offering car-sharing services in the business to consumer (B2C) system.

The purpose of the work was to present a review of the current status of car-sharing systems in Poland against the background of European systems.

## 2  Car-Sharing - Basic Information

Car-sharing, which is a solution that complements the offer of public transport in cities, is a concept which principle is based on the automated possibility of renting vehicles for a short period of time [18]. According to the assumptions of the system, it should be a form of renting a vehicle "for minutes" and constitute an alternative to owning a car in the city [19]. Steps related to renting a vehicle resemble the functioning of urban bicycle systems [20] and they are based on registration in a given system, choosing a vehicle and making a reservation, and then reaching the car, starting to drive, over-coming a specific route and returning the vehicle. The diagram of the next steps leading to car rental as part of car-sharing includes:

- downloading a mobile application of a given operator,
- registration in the operator's system,
- finding an interesting vehicle on the mobile map offered by the operator,
- reserving the vehicle,
- getting to the vehicle,
- starting the rental of the vehicle,
- ending the rent,
- payment of the fee.

Due to the different ways of returning the vehicle, there are three types of car-sharing: classic (so-called round-trip), one-way (so-called one-way) and free (so-called free-floating) [21–24].

Round-trip car-sharing is carried out when the vehicle is picked up and returned in exactly the same location [21, 22]. This form of car-sharing was one of the most basic services offered in the first world systems.

On the other hand, in one-way car-sharing, the vehicle collected in one place is to be returned at another point indicated by the operator of the given system [21, 22]. Such solutions usually mean the presence of a dedicated zone or car park where you should give away a rented car. The advantage of this solution is the parking space that will be waiting for the returned car. The disadvantage, however, is the need to return the car to the base, which does not necessarily have to be close to the destination that the user using the system wanted to reach.

The third type of car-sharing is the form that gives the most freedom when using the system. It is free-floating car-sharing. In this case, after renting the vehicle, the user has the greatest freedom as to the place of return of the vehicle, because he can leave it in any location within the city or cities where the operator operates [23, 24]. This solution gives a lot of freedom to the user due to the ability to go straight to the place he wants to reach. The disadvantage may be the lack of a parking space near the target.

Depending on the operator, the fleet of car-sharing systems can be vehicles with classic drive marked as fully thermic and vehicles with hybrid or electric drive marked as green [25, 26]. The target fleet, due to the issues of sustainable development of transport in cities, should be represented by cars from the group representing "green" type vehicles. Examples of hybrid and electric vehicles appearing in Polish car-sharing systems are presented in Fig. 1.

**Fig. 1.** Two examples of hybrid and electric vehicles used in one of pilotage Polish car-sharing systems

The history of car-sharing systems dates back to 1948, when the first such organization began to operate in Switzerland in Zürich [27]. In the years that followed, only individual practices appeared until the 1980s [28]. At that time, systems gradually began to appear in countries such as Italy, Germany, Austria, the Netherlands, England, Sweden, Denmark, Norway [28] and according to literature they functioned in a total of 450 Eutopean cities [28]. The greatest years of car-sharing prosperity have been the period since 2008, when after the economic crisis [29], the search for economical forms of mobility began. Currently, Germany is the leading operator on the European market. Car-sharing systems operate there in over 600 cities [30, 31] and the total number of registered users in 165 car-sharing systems operating in the country is about 2 million people [31].

## 3   Car-Sharing in Poland Overview

Car-sharing systems in Poland began their operations in 2016 in Krakow and Warsaw. At that time, two major service providers - Traficar and 4Mobility - appeared on the market. The service in Krakow began to be available to users from October 2016 [32]. The first 100 passenger cars of the B segment were available to the users [33]. Krakow's car-sharing was the first in Poland to apply a flexible system of renting vehicles, i.e. free-floating. Similarly to the Krakow system, in 2016, the 4Mobility operator started to operate in Warsaw. The main difference in the operation of the system was its type. The Warsaw system functioned on a round-trip basis and was characterized by base stations from which the car was picked up and to which it was to be returned after use [34]. Currently, after two years of operating on the market, Traficar operator's rentals are available in Kraków, Warsaw, Wrocław, Poznań, Tricity, Łódź, Bydgoszcz and Lublin as well as in the Upper Silesia conurbation in cities such as: Będzin, Bytom, Chorzów, Dąbrowa Górnicza, Gliwice, Jaworzno, Katowice, Mikołów, Mysłowice, Ruda Śląska, Siemianowice Śląskie, Sosnowiec, Tychy and Zabrze [35] as well as at the airport in Pyrzowice [35]. Along with the expansion of the system to include more cities, the fleet of vehicles has been modernized, and also equipped with additional delivery vehicles. In addition, the company started cooperation, among others with Orlen fuel stations, where special parking spaces have been devoted to Traficar vehicles. An example of parking spaces dedicated to car-sharing Traficar is presented in Fig. 2.

**Fig. 2.** The example of parking spaces dedicated to car-sharing

In turn, the services of the 4Mobility operator, currently outside of Warsaw, are also available in Poznań [34]. Currently, the operator operates in two types of car-sharing. Services are offered in basic car-sharing and free car-sharing [34]. Loan forms differ in terms of price. Additionally, as part of a base rental, it is possible to book a car up to 10 days in advance. As part of its service, the operator offers a diverse fleet of vehicles, ranging from C segment models to limousines belonging to the D premium segment [34].

In November 2017, the Vozilla operator began its operation in Wrocław, providing the first in Poland fully electric free-floating type car-sharing [36]. As part of the system, a fleet of 300 electric vehicles representing vehicles of the C-type and VAN-type cars was made available to users [36].

Next, new systems started to appear on the markets in Poznań (Click2Go, Easy-Share) or Warsaw (Panek Car-sharing).

In addition, pilot programs offering car-sharing services were conducted in selected cities. Such services include, among others, Warsaw Innogy programs or Drive Omni or the Lublin City Car operating in Lublin.

In total, there are currently six organized operators offering car-sharing services in Poland.

In addition to the services offered in organized car-sharing, i.e. provided by a specific operator, systems offering peer-to-peer (P2P) car-sharing services have started their activity in Polish cities. As part of this type of service, it is possible to share your vehicle with other users or to rent a vehicle from a private user through a dedicated car sharing system platform. The Locomto platform is a popular solution in Poland [37]. Due to the lack of a specific operator, many types of vehicles of various classes are available on the website. Rental rates are defined as an hourly basis, unlike organized operators. In addition, services often do not have vehicle insurance included in the rental price [37].

## 4  Car-Sharing in Poland - Analysis

In order to recognize the car-sharing market in Poland, the authors made an analysis based on operators providing organized car-sharing (B2B) services. Six operators operating in the country were considered for the analysis.

The obtained data show that organized car-sharing systems currently operate in 24 Polish cities. The rules for registration in systems are made in various ways and, depending on the operator, they can be more or less automated. In most Polish car-sharing systems, registration is done automatically by completing the online form. During registration, it is needed to provide personal details, PESEL number and to send a photo/scan of your driving license. In addition, one must also provide payment card details. Depending on the system, the operator can charge a 0.23 € registration fee. The required age of the user wishing to use the system is a minimum of 18 years. In the event of incorrect registration, the user must be present at the customer service office. Panek car-sharing is a unique method of registration on the Polish market. In this case, registration can be done in three ways: after completing the electronic form and connection to the operator's service via a video-conference or by visiting the branch or a traditional transfer of 0.23 € to the operator's account [38].

The method of renting a vehicle usually looks similar in the case of each system and consists of finding on the map, in a mobile application or on the map on the website of a given operator, an interesting vehicle and then booking it and reaching the vehicle. The time in which a user of Polish systems is to reach a vehicle is 15 or 20 min depending on the operator's requirements.

The fleet of vehicles used in Polish car-sharing systems is very diverse. Currently, 13 models of leading vehicles on the brands market are used in organized systems. The most frequent segment of vehicles used in organized systems is the C segment, constituting 43% of all vehicle models in Polish car-sharing. The "C" vehicle segment is a lower, average class of passenger cars, which is characterized by compact structure, providing however, relative comfort for 4 adults and moderately large luggage space. This type of segment is referred to as compact cars.

The vehicles used in the given systems do not repeat themselves in the systems of other operators, therefore, apart from the Toyota Yaris Hybrid model appearing in three organized car-sharing systems, it is impossible to determine the model of a vehicle with classic or electric power supply in Poland at the moment.

The diversity of vehicle segments in organized car-sharing has been presented in Fig. 3. The vehicles used in the given systems do not repeat themselves in the systems of other operators, therefore, apart from the Toyota Yaris Hybrid model appearing in three organized car-sharing systems, at this moment it is impossible to determine a leading model of a vehicle with classic or electric power supply in Poland. The diversity of vehicle segments in organized car-sharing is presented in Fig. 3.

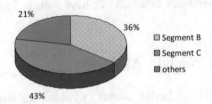

**Fig. 3.** Car-sharing vehicle segments in Poland

The size of the vehicle fleet in Polish car-sharing systems varies from 100 to 300 cars depending on the size of the city. However, these are mainly vehicles with classic drive. Hybrid vehicles in the largest number appear in the Warsaw system Panek car-sharing in the number of 300 cars, while electric cars reach their highest value in the Vozilla system in Wrocław in the amount of 200 units [36, 38]. The list of Polish car-sharing systems, due to the size of the fleet in individual cities, is presented in Table 1, whereas the comparison due to the models and vehicle segments in Table 2.

**Table 1.** List of vehicle fleets in Polish car-sharing systems (Source: author's own collaboration based on: [32–36, 38, 40]).

| Operator | City | Fleet size |
|---|---|---|
| Traficar | Warsaw | 300 |
| | Kraków | 300 |
| | Wrocław | 150 |
| | Poznań | 150 |
| | Tricity | 300 |
| | Silesia and Dąbrowa Basin | 300 |
| 4Mobility | Warsaw | 300 |
| | Poznań | 30 |
| Panek car-sharing | Warsaw | 300 |
| Vozilla | Wrocław | 300 |
| Clock2go | Poznań | 100 |
| EasyShare | Poznań | 200 |

**Table 2.** List of vehicle fleets in Polish car-sharing systems due to vehicle models and segments (Source: author's own collaboration based on: [32–36, 38, 40]).

| Operator | Vehicle model | Vehicle segment |
|---|---|---|
| Traficar | Renault Clio | B |
| | Opel Corsa | B |
| | Renault Kangoo | VAN |
| 4Mobility | Hyundai i30 | C |
| | Mini One | C |
| | BMW 1 | C |
| | Audi A3 | C |
| | BMW 3 | D |
| | Audi Q3 | Crossover C |
| | BMW i3 | D |
| Panek car-sharing | Toyota Yaris Hybrid | B |
| Vozilla | Nissan Leaf | C |
| | Nissan eNV200 | VAN |
| Clock2go | Toyota Yaris Hybrid | B |
| EasyShare | Toyota Yaris Hybrid | B |

The predominant type of car-sharing system is the model of the possibility of leaving the vehicle anywhere in the city - free-floating. And the analysis shows that the system is characterized by 83.3% of all operators operating in Poland. The list of car-sharing systems operating in Poland due to the type of system and the city of operation is presented in Table 3.

**Table 3.** List of Polish car-sharing systems due to system type (Source: author's own collaboration based on: [32–36, 38, 40]).

| Operator | System type | Cities |
|---|---|---|
| Traficar | Free-floating | Warsaw, Kraków, Wrocław, Poznań, Tricity, Silesia and Dąbrowa Basin Łódź Bydgoszcz, Lublin |
| 4Mobility | Free-floating, station based | Warsaw Poznań |
| Panek car-sharing | Free-floating | Warsaw |
| Vozilla | Free-floating | Wrocław |
| Clock2go | Free-floating | Poznań |
| EasyShare | Free-floating | Poznań |

Due to the cost of car-sharing, in most Polish systems, in contrast to European systems, the cost of the service consists of travel time and distance and a possible parking charge (in the case of closing the vehicle and leaving it "borrowed" on the user's account, and not returned - called stop over).

Depending on the system, the prices for the travel time of the rented vehicle range between PLN 0.25/min–1 PLN/min, which is on average 0.06 €/min–0.23 €/min depending on the class of the vehicle rented. For this, the distance costs are 0.65 PLN/km–1.20 PLN/km, which corresponds to an average of 0.15 €–0.28 €. However, the vehicle stop-over fee is on average 0.10 PLN, which corresponds to 0.02 €. In most systems, service packages are not offered.

The list of Polish car-sharing systems due to the costs of vehicle use is presented in Table 4.

**Table 4.** List of Polish car-sharing systems due to usage costs (Source: author's own collaboration based on: [32–36, 38–40]).

| Operator | Clearance type | Amount of fee per km | Amount of fee per minute | Stop-over fee |
|---|---|---|---|---|
| Traficar | Time-kilometer fee | 0.19 €/km | 0.12 €/min | 0.02 €/min |
| 4Mobility | Time-kilometer fee | 0.19–0.21 €/km | 0.08–0.20 €/min | 0.03–0.04 €/min |
| Panek car-sharing | Time-kilometer fee | 0.15 €/km | 0.12 €/min | 0.02 €/min |
| Vozilla | Time | - | 0.21 €/min | 0.02 €/min |
| Clock2go | Time-kilometer fee | 0.19 €/km | 0.12 €/min | 0.02 €/min |
| EasyShare | Time-kilometer fee | 0.19 €/km | 0.12 €/min | 0.02 €/min |

## 5  Summary

In summary, based on the analysis, it can be concluded that the implementation of car-sharing services in Poland has gained in popularity. It has gained popularity since the appearance of the first systems on the Polish market in 2016. Despite this, in relation to the systems functioning on other European markets, Poland is at the beginning to the development of short-term rental vehicles.

In the current state of car-sharing in Poland, it is therefore one of the possibilities to support sustainable mobility. However, in order to fully achieve the intended objectives, due to the policy of sustainable transport one should focus, among others, on:

- on the aspect of the fleet of vehicles and their modernization for electric cars in line with the trend of electro-mobility,
- increasing the number of available vehicles,
- relocation of vehicles between strategic places and neighboring cities,
- normalizing the pricing policy of services based on Western trends of rental fees,
- applying the option of purchasing a package of services,
- sharing mobility promotions.

Then, car-sharing has a chance to become a tool for achieving sustainable urban mobility.

# References

1. Pinderhughes, R.: Alternative Urban Futures: Planning for Sustainable Development in Cities Throught the World. Rowman & Littlefield Publishers, New York (2004)
2. Taniguchi, E., Thompson, R.G., Yamada, T.: Recent trends and innovations in modelling city logistics. Procedia – Soc. Behav. Sci. **125**, 4–14 (2014)
3. Staniek, M.: Moulding of travelling behaviour patterns entailing the condition of road infrastructure. In: Macioszek, E., Sierpiński, G. (eds.) Contemporary Challenges of Transport Systems and Traffic Engineering. LNNS, vol. 2, pp. 181–191. Springer, Cham (2017)
4. Okraszewska, R., Nosal, K., Sierpiński, G.: The role of the Polish Universities in shaping a new mobility culture - assumptions, conditions, experience. Case Study of Gdansk University of Technology, Cracow University of Technology and Silesian University of Technology. In: Proceedings of ICERI2014 Conference, pp. 2971–2979. ICERI Press, Seville (2014)
5. Sierpiński, G.: Ocena Systemu Transportowego oraz Preferencje Osób Podróżujących, Jako Wsparcie Kształtowania Podziału Zadań Przewozowych - Studium Przypadku dla Konurbacji Górnośląskiej. Prace Naukowe Politechniki Warszawskiej, Transp. **111**, 487–499 (2016)
6. Staniek, M.: Detection of cracks in asphalt pavement during road inspection processes. Sci. J. Silesian Univ. Technol. Ser. Transp. **96**, 175–184 (2017)
7. Macioszek, E., Czerniakowski, M.: Safety-related changes introduced on T. Kościuszki and Królowej Jadwigi Streets in Dąbrowa Górnicza between 2006 and 2015. Sci. J. Silesian Univ. Technol. Ser. Transp. **96**, 95–104 (2017)
8. Turoń, K., Czech, P., Juzek, M.: The concept of Walkable City as an alternative form of urban mobility. Sci. J. Silesian Univ. Technol. Ser. Transp. **95**, 223–230 (2017)
9. Macioszek, E., Lach, D.: Analysis of the results of general traffic measurements in West Pomeranian Voivodeship over the years 2005–2015. Sci. J. Silesian Univ. Technol. Ser. Transp. **97**, 93–104 (2017)
10. Staniek, M.: Stereo vision method application to road inspection. Baltic J. Road Bridge Eng. **12**(1), 38–47 (2017)
11. Okraszewska, R., Romanowska, A., Wołek, M., Oskarbski, J., Birr, K., Jamroz, K.: Integration of a multilevel transport system model into sustainable urban mobility planning. Sustainability **10**(2), 1–20 (2018)
12. Macioszek, E.: First and last mile delivery - problems and issues. In: Sierpiński, G. (ed.) Advanced Solutions of Transport Systems for Growing Mobility. AISC, vol. 631, pp. 147–154. Springer, Cham (2018)
13. Jacyna-Gołda, I.: Decision-making model for supporting supply chain efficiency evaluation. Arch. Transp. **33**(1), 17–31 (2015)
14. Turoń, K., Golba, D., Czech, P.: The analysis of progress CSR good practices areas in logistic companies based on reports "Responsible Business in Poland. Good Practices" in 2010–2014. Sci. J. Silesian Univ. Technol. Ser. Transp. **89**, 163–171 (2015)
15. Shaheen, S., Chan, N., Rayle, L.: Ridesourcing's Impact and Role in Urban Transportation. https://www.accessmagazine.org/wp-content/uploads/sites/7/2017/05/Shaheen-Rayle-and-Chan-Access-Spring-2017.pdf

16. Hebel, K., Wołek, M.: Change trends in the use of passenger cars on urban trips: car-pooling in Gdynia. Sci. J. Silesian Univ. Technol. Ser. Transp. **96**, 37–47 (2017)
17. Ferrero, F., Perboli, G., Rosano, M., Vesco, A.: Car-sharing services: an annotated review. Sustain. Cities Soc. **37**, 501–518 (2018)
18. European Commission: Communication from the Commission to the European Parliament, the Council, the European Economic and Social Committee and the Committee of the Regions, Online Platforms and the Digital Single Market, Opportunities and Challenges for Europe. European Commission, Brussels (2016)
19. Britton, E., World Carshare Associates: Carsharing 2000. Sustainable Transport's Missing Link. The Journal of World Transport Policy & Practice. Eco-Logica, Lancaster (2000)
20. Czech, P., Turoń, K., Urbańczyk, R.: Bike-sharing as an element of integrated urban transport system. In: Sierpiński, G. (ed.) Advanced Solutions of Transport Systems for Growing Mobility. AISC, vol. 631, pp. 103–111. Springer, Cham (2018)
21. Shaheen, S.A., Chan, N.D., Micheaux, H.: One-way carsharing's evolution and operator perspectives from the Americas. Transportation **42**(3), 519–536 (2015)
22. Ferrero, F., Perboli, G., Musso, S., Vesco, A.: Business Models and Tariff Simulation in Car-Sharing Services. https://pdfs.semanticscholar.org/7beb/a2dbf464935be7a2d452702069e35550922ed.pdf
23. Nourinejad, M., Roorda, M.: Carsharing operations policies: a comparison between one-way and two-way systems. Transportation **42**(3), 497–518 (2015)
24. Ciari, F., Bock, B., Balmer, M.: Modeling station-based an free-floating carsharing demand: a test case study for Berlin, Germany. Emerging and innovative public transport and technologies. Transportation Research Board of the National Academies, Washington (2014)
25. Firnkorn, J., Müller, M.: What will be the environmental effects of new free-floating car-sharing systems. The case of Car2go in Ulm. Ecol. Econ. **70**(8), 1519–1528 (2011)
26. Sierpiński, G.: Model of incentives for changes of the modal split of traffic towards electric personal cars. In: Mikulski, J. (ed.) Telematics - Support for Transport. CCIS, vol. 471, pp. 450–460. Springer, Heidelberg (2014)
27. Harms, S., Truffer, B.: The Emergence of A Nationwide Carsharing Co-operative in Switzerland. Research report, EAWAG, Dübendorf (1998)
28. Doherty, M.J., Sparrow, F.T., Sinha, K.C.: Public use of autos: mobility enterprise project. ASCE J. Transp. Eng. **113**(1), 84–94 (1987)
29. Ziobrowska, J.: Sharing Economy Jako Nowy Trend Konsumencki. Własność w Prawie i Gospodarce. http://www.bibliotekacyfrowa.pl/Content/79527/Wlasnosc_w_prawie_i_gospodarce.pdf
30. Fleet Europe: Germany Enacts Car-sharing Law. https://www.fleeteurope.com/en/news/germany-enacts-car-sharing=law
31. Current Data and Data on the CarSharing Service in Germany. https://carsharing.de/alles-ueber-carsharing/carsharing-zahlen/aktuelle-zahlen-daten-zum-carsharing-deutschland
32. Polish Parking Portal - Polska Parkuje. http://polskaparkuje.pl/2016/10/05/ruszyl_traficar
33. Polish Parking Portal - Traficar Information - Polska Parkuje. http://polskaparkuje.pl/wp-content/uploads/2016/10/Traficar_prezentacja-na-konferencje%CC%ezentacja-na-konferencje%CC%A8_20161012.pdf
34. Mobility Operator. http://4mobility.pl/
35. Traficar Operator. https://www.traficar.pl/
36. Vozilla Operator. https://www.vozilla.pl/
37. Locomoto Polish Car-Sharing Platforms. http://locomoto.pl/
38. Panek Car-Sharing. https://panekcs.pl
39. Click2go Car-Sharing. https://www.click2go.pl/
40. Easyshare Car-Sharing. https://easyshare.pl/

# Estimation of the Number of Threatened People in a Case of Fire in Road Tunnels

Aleksander Król[1](✉) and Małgorzata Król[2]

[1] Faculty of Transport, Silesian University of Technology, Katowice, Poland
aleksander.krol@polsl.pl
[2] Faculty of Energy and Environmental Engineering,
Silesian University of Technology, Gliwice, Poland
gosia.krol@polsl.pl

**Abstract.** Factors influencing the conditions in a road tunnel in a case of fire were considered. The specifics of road tunnels as parts of the transportation network from the safety point of view was discussed. A brief overview on tunnel emergency systems and evacuation process was presented. Selected road tunnels in Poland were taken into account and a number of scenarios of traffic accidents with fire were detailed described. Some observations from carried-out full-scale hot smoke test were used. As a result of each such case study the estimated number of threatened tunnel users was obtained. Conclusion section contains some recommendations which would lead to increase the safety level in road tunnels.

**Keywords:** Road tunnel · Fire outbreak · Evacuation · Traffic safety

## 1 Introduction

Road tunnels have become an indispensable part of the road network. They can be found in cities where they facilitate urban transport as well as in mountain areas due to avoid detours. Tunnels are equipped with a number of systems to ensure the safe use and in case of fire, to support the evacuation. It concerns especially the tunnels longer than 400 m, the detailed standards depend on tunnel type, its localisation and traffic intensity [1]. Table 1 presents road tunnels in Poland longer than 400 m.

A road tunnel is a part of the road network, but the issues of people safety have to be considered in the different way than for other roads. In general, traffic accidents in tunnels happen less often because of lack of crossroads, the commonly limited speed, the independency on weather conditions and the lack of pedestrians. On the other side, if an accident happens the consequences are commonly much more severe [2]. The threat results mainly from the limited space and an impeded evacuation and access of rescue teams. The situation becomes significantly worse in a case of fire outbreak. In such case the main dangerous factor is smoke, which is a mixture of air and toxic gases. Smoke poisoning can quickly lead to unconsciousness and death. In addition smoke worsens the visibility, what can disturb the evacuation. In the space close to the fire the smoke is hot and can cause burns.

© Springer Nature Switzerland AG 2019
E. Macioszek and G. Sierpiński (Eds.): Directions of Development of Transport
Networks and Traffic Engineering, LNNS 51, pp. 27–40, 2019.
https://doi.org/10.1007/978-3-319-98615-9_3

**Table 1.** Road tunnels in Poland.

| No | Name | Location | Length |
|---|---|---|---|
| 1 | Tunnel under Martwa Wisła river | Gdańsk | 1377 m |
| 2 | Tunnel of Wisłostrada | Warszawa | 930 m (S)/889 m (N) |
| 3 | Tunnel Emilia | Laliki, expressway S1 | 678 m |
| 4 | Katowicki Tunnel | Katowice | 657 m (S)/650 m (N) |
| 5 | Tunnel of diametrical road in Gliwice | Gliwice | 493 m |

As was mentioned above, road accidents and fires in tunnels occur rarely, but the growing number of tunnels in the World (also in Poland) and the growing number of cars cause they will be inevitable. A study carried out in France [3] suggested that:

- there will be one or two serious fires per one kilometre of tunnel length for every 100 million passenger cars passing a tunnel,
- there will be about eight fires per one kilometre of tunnel length for every 100 million trucks passing a tunnel, one of them will be so serious to cause people life threat and large material damage.

This forecast is obviously based on analysis of the past road accidents and fires in tunnels. Table 2 presents the most serious disasters in road tunnels.

**Table 2.** Selected tunnel accidents with fire.

| Tunnel | Country | Year | Reason | Casaulties |
|---|---|---|---|---|
| Gothard | Switzerland | 2001 | Truck collision | 11 |
| Gleinalmtunnel | Austria | 2001 | Vehicle collision | 5 |
| Tauern | Austria | 1999 | Vehicle collision | 12 |
| Mont Blanc | France - Italy | 1999 | Cigarette butt in the air filter of a truck | 39 |
| Isola delle Femmine | Italy | 1996 | 2 vehicle collisions | 5 |
| Pecorile | Italy | 1983 | Vehicle collision | 8 |
| Holland | NY USA | 1949 | Fall of the cargo from a truck | 66 injured |

A fire in a road tunnel commonly develops very quickly, therefore the conditions inside are getting worse and worse in a very short time. It is even possible that the fire "jumps" onto a distant car [4]. According to the real scale burning tests the heat release rate is about 4000 kW for a passenger car fire and about 30000 kW for a truck fire, the time needed for full fire development is of order of few minutes [5]. So, rescue teams are not able to undertake the rescue action immediately. In such situation, the self-rescue action of threatened people is the only way to save. All tunnels systems must support this phase of rescue action. Certainly it concerns the tunnel ventilation system as well [6].

The paper presents a few case studies considering the fire outbreak in different real road tunnels. The fire location, tunnel characteristic and the volume of road traffic were taken into account. In each case a scenario of an accident and then the fire development was shown.

## 2  Factors Influencing the Number of Threatened People and the Evacuation Process

The number of threatened people depends on traffic intensity i.e. number of cars which get stuck in a tunnel, the location of a fire, the operation of the tunnel emergency systems and the possibilities of self-evacuation. All these issues are briefly discussed below.

### 2.1  Systems Supporting Self-evacuation in Road Tunnels

Systems in a road tunnel have to meet a series of technical requirements set out in the Ordinance of the Minister of Transport and Maritime Economy of 30th May 2000 on technical conditions to be met by traffic engineering objects and their location [7]. In case of fire the most important challenge is to remove the hot smoke and the other toxic products of combustion and to keep as large volume free of smoke as possible.

There are three solutions applied for road tunnel ventilation systems: longitudinal systems, semi-transverse systems and transverse systems. Longitudinal systems are designed for tunnels shorter than 1000 m, air exchange goes just by tunnel portals and the air flow is forced by axial fans. Transverse systems include additional ducts equipped with fans which provide fresh air supply or smoke extraction in selected locations of the tunnel.

Regardless of the ventilation system mounted there is a general rule which determines the mode of system operation: the smoke must be removed in such a way to avoid the smokiness of most volume of the tunnel. In a case of longitudinal systems it means that the smoke must be removed by the nearest tunnel portal (Fig. 1).

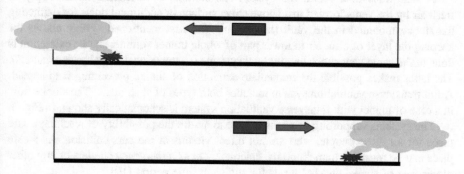

**Fig. 1.** Direction of forced airflow in dependence on the location of fire outbreak (longitudinal ventilation system)

If the road traffic in a given tunnel is unidirectional the fans should be switched on accordingly to the traffic direction, unless the fire is located quite close to the entrance portal. Additionally, if the traffic is congested attention must be paid to safety of the drivers, who have just passed the accident place, because they are not able to leave quickly. The natural draught caused by tunnel inclination, temperature difference or wind should be also taken into account [8]. Therefore, there is an emergency operation pattern established for each tunnel, it determines the fans to be switched on, and their work direction in dependency on location of fire outbreak and intensity natural draught.

Full transverse ventilation as the only solution for ventilation in the tunnel is rare. Most often it concerns tunnels over 2000 m. In other tunnels the longitudinal ventilation is met. For the longitudinal ventilation systems the most important matter is whether a system is able to keep so called critical velocity of air [9]. The importance of critical velocity is shown in Fig. 2. The air flow forced by fans must be able to prevent the smoke upstream movement, but the air velocity must not be too high, because it could disturb the stratification of air and smoke layers.

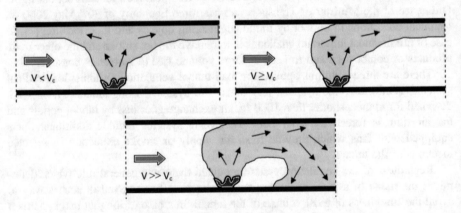

**Fig. 2.** Significance of critical velocity

A semi-transverse ventilation system can consists of additional ducts supplying fresh air by the vents located just above street surface or additional ducts for extracting the smoke mounted in the vault (less often). The first mentioned system allows for keeping the layer of clear air in lower part of whole tunnel volume, smoke extraction is done in the same way as for longitudinal systems (so the critical velocity also matters). The latter makes possible the immediate extraction of smoke preventing it to spread. A full transverse ventilation system includes both types of such ducts. The smoke flow in a case of tunnel with transverse ventilation system is schematically shown in Fig. 3.

The systems supporting evacuation have to assure the possibility of reaching a safe place for all those people, who are not direct victims of the cars collision. As a safe place can be regarded tunnel portals, emergency exits, emergency shelters or any other places free of smoke and hot gases for sufficient time period [10].

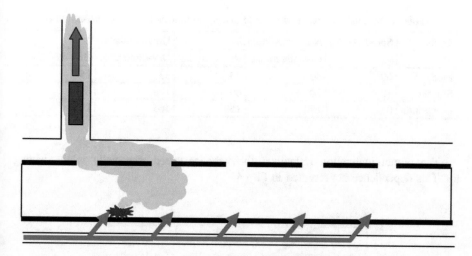

**Fig. 3.** Operation of full transverse ventilation system

According to the standard PD 7974-6 [11] the critical conditions which make the evacuation possible are as follows:

- temperature cannot exceed 68 °C,
- visibility cannot be lower than 10 m,
- the thermal radiation flux density is 2.5 kW/m$^2$ for a duration of exposure longer than 30 s,
- oxygen concentration cannot be lower than 15%.

If the self-evacuation is to be succeed the time of exposure to the fire threats should be minimized, therefore the distance to the nearest emergency exit cannot be too long. Emergency exits commonly lead to the parallel evacuation tunnel (in a case of bidirectional road traffic) or to the second tube (in a case of unidirectional road traffic). As a rule each country adopted different establishing, which regulates the issue of the admissible distance to the nearest safe place. And so, the maximum distance between emergency exits is established as no greater than 500 according to the EP Directive 2004/54/EC [12] or as no greater than 300 according to the German Standard RABT [13].

Self-evacuation should be supported by clear emergency marking and audio messages, which prompt the drivers and passengers to leave cars and direct them properly towards safe places.

## 2.2 Estimation of Traffic Volume

The number of people in threatened zone depends on the number of cars in traffic flow. The traffic flow intensity depends mainly on road type and day time, but tunnels are specific facilities even for a reason of limited traffic velocity. Some studies were done and there are the standard maximum values of traffic density adopted (Table 3).

**Table 3.** Standardised maximum traffic density in tunnels [vehicles/lane/km] [14].

| Traffic | Speed [km/h] | Non-urban tunnels | | Urban tunnels | |
|---------|-------------|-------------------|------------------|------------------|------------------|
| | | One-directional | Bi-directional | One-directional | Bi-directional |
| Free | 60 | 30 | 23 | 33 | 25 |
| Slowed | 10 | 70 | 60 | 100 | 85 |
| Congestion | 0 | 150 | 150 | 165 | 165 |

The tunnel inclination influences the vehicles speed, especially while driving up. This dependence is presented in Fig. 4.

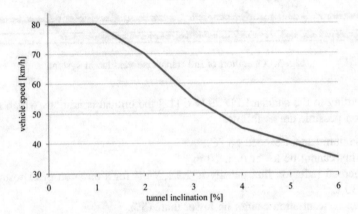

**Fig. 4.** Dependence of vehicles speed on tunnel inclination [14]

## 2.3 Overall Evacuation Schedule

The time factor is crucial when considering the possibility of self-evacuation in a case of fire in a road tunnel. Formally it can be expressed as follows:

$$T_{ASET} - T_{RSET} \geq 0 \tag{1}$$

where:

$T_{ASET}$ - available safe escape time,
$T_{RSET}$ - required safe escape time.

The determination of available safe escape time ($T_{ASET}$) requires the studies on variable conditions in a tunnel in case of fire outbreak. The mentioned above critical conditions must be assured to give people the chance to survive. It can include using the standards, the results of full-scale or model experiments or numerical analyses. In turn, the required safe escape time ($T_{RSET}$) is a sum of time periods, which are related to successive stages of evacuation process:

$$T_{RSET} = t_{det} + t_{alm} + t_{rcg} + t_{rct} + t_{mov} \ [min] \tag{2}$$

where:

$t_{det}$ - time of fire detection,
$t_{alm}$ - time of alarm,
$t_{rcg}$ - time needed for recognition and situation comprehension,
$t_{rct}$ - time for reaction - selection of escape route,
$t_{mov}$ - time of movement to the chosen safe place.

The above relation can be presented graphically, what is shown in Fig. 5 [15].

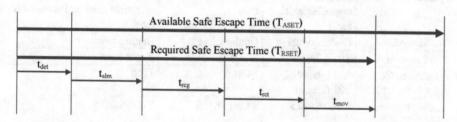

**Fig. 5.** Components of required safe escape time and relation to available safe escape time

The relations shown in Fig. 5 determine the ways which can be used to increase the reserve of time allowing for safe evacuation:

- increase of available safe escape time by accurately design ventilation systems and other tunnel systems,
- decrease of time needed for precise fire detection,
- immediate, unequivocal and clear emergency message,
- clear marking of escape routes,
- short distance to nearest safe place,
- support for evacuation of disabled persons,
- wide awareness campaigns of the principles of proper behaviour in the event of a fire hazard.

# 3   Results

Possible consequences of a traffic accident in three real road tunnels were considered. For each tunnel three different fire locations and different traffic conditions were assumed. The brief description of the selected tunnels is presented in Table 4. The data concerning traffic volume and structure were adopted according to the results of General Traffic Measurement 2015 [16].

Basing on real data on similar cases [15, 17] the following numbers of passengers per vehicle were assumed: 2 per a car, 1 per a truck and 15 per a bus.

**Table 4.** Short description of selected tunnels.

| Tunnel | Laliki Tunnel | Katowicki Tunnel | Tunnel under Vistula river |
|---|---|---|---|
| Location | Non-urban | Urban | Urban |
| Structure | One tube | Two tubes | Two tubes |
| Length [m] | 678 | 650 | 1377 |
| Shape | Vault | Rectangle | Circle |
| Orientation | NE-SW | W-E | NW-SE |
| Inclination [%] | 4 towards NE | 0 | 3 and 4 from the deepest point towards portals |
| Ventilation system | Longitudinal (5 pairs of reversible axial fans) | Longitudinal (14 axial fans in each tube) | Semi-transverse (vents supplying air, 11 single reversible axial jet fans in each tube) |
| Traffic mode | Bi-directional | Unidirectional | Unidirectional |
| Road structure | 2 × 1 lane | 2 × 3 lanes | 2 × 2 lanes |
| Average traffic volume [veh/h] | 700 (planned), 200 (actual) | 1400 | 1100 |
| Traffic structure [%] cars/trucks/buses | 83.6/15.9/0.5 | 82.9/16.6/0.5 | 73.0/26.4/0.6 |
| Rush hour traffic | Free | Congested/slowed | Congested/slowed |
| Speed limit [km/h] | 80 | 70 | 70/50 |

The fundamental relation governing the road traffic was also used in following consideration (for single lane):

$$Q = k \cdot v \tag{3}$$

where:

$Q$ - traffic intensity [veh/h],
$k$ - traffic density [veh/km],
$v$ - traffic velocity [km/h].

Having in mind all mentioned above issues and relations the number of cars which get stuck and then the number of threatened people were estimated. In all discussed cases the vehicles which were just in tunnel and those, which entered before triggering the alarm and were not able to leave were regarded as got stuck. The passengers in such vehicles were regarded as threatened if they were in the fully smoky part of the tunnel. It was assumed that for the free driving the vehicles which passed the fire location were able to leave the tunnel. In a case of slowed traffic or congestion also these cars were regarded as threatened (the presented estimations did not take into account the time needed for smoke spreading).

## 3.1   Laliki Tunnel

Figure 6 presents the sketch of three assumed fire locations. As the actual or even planned traffic intensities (respectively 200 and 700 veh/h) are not too high, the free driving can be adopted as the most probably traffic condition. Eventually the value of the traffic intensity equal to 500 veh/h was adopted in below considerations.

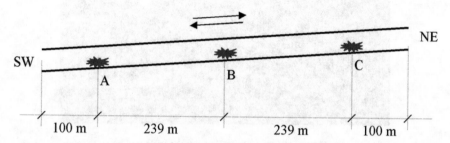

**Fig. 6.**  Assumed fire locations for Laliki Tunnel (traffic direction is marked)

According to the emergency pattern of fire ventilation [18] the direction of fans operation after fire detection should be SW → NE for cases B and C and NE → SW for case A. The time needed to arouse the fire alarm is about 1 min [19]. After then the red lights at tunnel portals will forbid the entrance. The expected scenarios of fire accidents are shown in Table 5.

**Table 5.**  Scenarios for fire accident in Laliki Tunnel.

| Case | Driving direction | Traffic | $Q$ [veh/h] | $k$ [veh/km] | $v$ [km/h] | No of cars got stuck | | | No of threatened people |
|------|------|------|------|------|------|------|------|------|------|
| | | | | | | Total | Due to fire | In smoke | |
| A | SW → NE | Free | 500 | 10 | 50 | 9 | 9 | 9 | 18 |
| ← | NE → SW | | 500 | 7 | 70 | 12 | 12 | 0 | |
| B | SW → NE | Free | 500 | 10 | 50 | 11 | 11 | 0 | 20 |
| → | NE → SW | | 500 | 7 | 70 | 10 | 10 | 10 | |
| C | SW → NE | Free | 500 | 10 | 50 | 14 | 14 | 0 | 16 |
| → | NE → SW | | 500 | 7 | 70 | 8 | 8 | 0 | |

The above estimation took into account that the tunnel was equipped with longitudinal ventilation system, so in some cases the leeward side would be full of smoke. Figure 7 presents the smoke fully filling the tunnel during hot smoke tests.

**Fig. 7.** Smoke on leeward during hot smoke tests in Laliki Tunnel

## 3.2 Katowicki Tunnel

Figure 8 presents the sketch of three assumed fire locations. The tunnel consists of two unidirectional tubes, so only one of them was analysed. The southern tube leading from W towards E was taken into account. The traffic intensity is very high at rush hour, thus slowed and congested traffic conditions were examined.

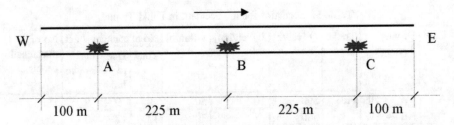

**Fig. 8.** Assumed fire locations for Katowicki Tunnel (traffic direction is marked)

The fans should be switched on accordingly to the traffic direction (W → E for cases B and C), unless the fire is located quite close to the W portal (case A). The time elapsed between the fire outbreak and detection and alarm triggering was assumed to be 1 min too. The expected scenarios of fire accidents are shown in Table 6.

This tunnel is also equipped with longitudinal ventilation system, so as previously there is a danger that the leeward side would be fully smoky. However, the number of cars which were regarded as totally engulfed by smoke in case of slowed traffic was estimated with a great excess. This was because a significant part of such cars would be able to leave the danger zone.

**Table 6.** Scenarios for fire accident in Katowicki Tunnel.

| Case | Driving direction | Traffic | $Q$ [veh/h] | $k$ [veh/km] | $v$ [km/h] | No of cars got stuck | | | No of threatened people |
|------|------|------|------|------|------|------|------|------|------|
| | | | | | | Total | Due to fire | In smoke | |
| A | W → N | Slowed | 3 × 1000 | 3 × 100 | 10 | 80 | 80 | 80 | 146 |
| ← | | Congested | 0 | 3 × 165 | 0 | 321 | 50 | 50 | 91 |
| B | | Slowed | 3 × 1000 | 3 × 100 | 10 | 150 | 150 | 100 | 183 |
| → | | Congested | 0 | 3 × 165 | 0 | 321 | 160 | 160 | 293 |
| C | | Slowed | 3 × 1000 | 3 × 100 | 10 | 210 | 210 | 30 | 55 |
| → | | Congested | 0 | 3 × 165 | 0 | 321 | 271 | 49 | 87 |

## 3.3    Tunnel Under Martwa Wisła River

Figure 9 presents the sketch of three assumed fire locations. The tunnel also consists of two unidirectional tubes, so only one of them was analysed. The southern tube leading from NW towards SE was taken into account. The traffic intensity is very high at rush hour, thus slowed and congested traffic conditions were examined.

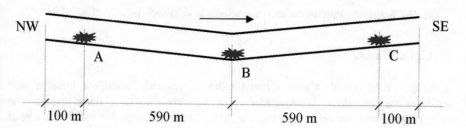

**Fig. 9.** Assumed fire locations for Tunnel under Martwa Wisła river (traffic direction is marked)

The fans should be switched on accordingly to the traffic direction (NW → SE for cases B and C), unless the fire is located quite close to the NW portal (case A). The time elapsed between the fire outbreak and detection and alarm triggering was assumed to be 1 min too. The expected scenarios of fire accidents are shown in Table 7.

**Table 7.** Scenarios for fire accident in Tunnel under Martwa Wisła river.

| Case | Driving direction | Traffic | $Q$ [veh/h] | $k$ [veh/km] | $v$ [km/h] | No of cars got stuck | | | No of threatened people |
|------|------|------|------|------|------|------|------|------|------|
| | | | | | | Total | Due to fire | In smoke | |
| A | NW → SE | Slowed | 2 × 1000 | 2 × 100 | 10 | 53 | 53 | 53 | 97 |
| ← | | Congested | 0 | 2 × 165 | 0 | 455 | 33 | 33 | 60 |
| B | | Slowed | 2 × 1000 | 2 × 100 | 10 | 171 | 171 | 20 | 36 |
| → | | Congested | 0 | 2 × 165 | 0 | 455 | 227 | 33 | 60 |
| C | | Slowed | 2 × 1000 | 2 × 100 | 10 | 288 | 288 | 20 | 36 |
| → | | Congested | 0 | 2 × 165 | 0 | 455 | 422 | 33 | 60 |

This tunnel is equipped with semi-transverse ventilation system with ducts supplying fresh air through the vents just above the street. Axial fans force the smoke to flow towards one of portals, so the leeward side is partially smoky. The lower part of leeward side is free of smoke, except the region in the vicinity of the fire: about 100 m is full of smoke (Fig. 10). Just this value was applied to estimate the number of cars in smoky region. And again as before, probably most of cars in this region would be able to leave the danger zone.

**Fig. 10.** Smoke on leeward during hot smoke tests in Tunnel under Martwa Wisła river

## 4   Conclusions

A fire in a road tunnel is great threat for the drivers and passengers. Despite such accidents are rare, their consequences would be very serious. Having in mind the increasing number of road tunnels one should be aware that such fire accident in a road tunnel will rather occur in the future. As was proved the number of people whose life and health would be threatened in such accident would be large.

As was shown the number of threatened people depended on the number of cars got stuck, fire location and operation mode of ventilation system. The presented results were supported by the observations made during hot smoke tests carried out in real road tunnels. The direct victims of a road accident were not taken into account, therefore the threat coming from spreading smoke was regarded as the main one. The attention was focused on the people who were in the cars in tunnel at the moment of accident and were not able to leave and also in the cars, which entered before the tunnel was closed. From this point of view the traffic intensity in a tunnel is crucial, then goes the issue whether the traffic is unidirectional or bidirectional. In a case of congestion, all cars in tunnel are not able to leave it and all passengers are exposed to the toxic smoke, especially those on leeward. Similarly, for bidirectional traffic there are cars got stuck at both sides of the place of road accident and fire.

The exact number of threatened people is unpredictable, as it strongly depends on coincidence of a set of random factors. The above provided scenarios described some typical cases and allowed to realize the importance of the problem.

The most dangerous conditions could emerge when in case of longitudinal ventilation system the critical velocity is significantly exceeded and the stratification of smoke and air layers is disturbed. Then there is a danger that leeward part of the tunnel would be full of smoke. A semi-transverse ventilation system could stave off this problem by supplying fresh air to the lower part of tunnel. On the other hand such solution significantly increases building costs and is not necessary in shorter tunnels. So it is clear why tunnel ventilation and safety system have to be designed carefully.

Furthermore, the trainings of rescue teams have to be carried out often. They should include the whole scenario including the period between the emergency fire detection signal and briefing after the completion of rescue action. It would also advisable to establish the periodical audits of tunnel safety systems, which would examine not only the efficiency of each single system, but first of all their cooperation in case of fire would be assessed. The next important issue is an awareness campaign of the principles of proper behaviour in the event of a fire hazard among tunnel users.

# References

1. Verein Deutscher Ingenieure: Ventilation Plants for Road Tunnels. VDI 6029. Verein Deutscher, Dusseldorf (2000)
2. Beard, A., Carvel, R.: The Handbook of Tunnel Fire Safety. Thomas Telford Ltd., London (2005)
3. Chojnacki, K., Fabryczewska, A.: Bezpieczeństwo Pożarowe w Tunelach. Górnictwo i Geoinżynieria 3(1), 145–156 (2005)
4. Massachusetts Highway Department and Federal Highway Administration: Memorial Tunnel Fire Ventilation Test Program. Test Report. Technical report, Bachtel, Massachusetts (1995)
5. Jannsens, M.: Development of a Database of Full-Scale Calorimeter Tests of Motor Vehicle Burns. Southwest Research Institute, San Antonio (2008)
6. Kumar, S.: Recent Achievements in Modelling the Transport of Smoke and Toxic Gases in Tunnel Fires. https://about.ita-aites.org/.../363_2ff5fa1595490b498a0b095cbc3
7. Ministerstwo Transportu i Gospodarki Morskiej: Rozporządzenie Ministra Transportu i Gospodarki Morskiej z Dnia 30 maja 2000 r. w Sprawie Warunków Technicznych, Jakim Powinny Odpowiadać Drogowe Obiekty Inżynierskie i Ich Usytuowanie (Ze Zmianami z Dnia 29 maja 2012). Dz.U.2000.63.735. Ministerstwo Transportu i Gospodarki Morskiej, Warsaw (2000)
8. Król, A., Król, M.: Impact of the factors determining the natural stack effect on the safety conditions in a road tunnel. In: Macioszek, E., Sierpiński, G. (eds.) Recent Advances in Traffic Engineering for Transport Networks and Systems. LNNS, vol. 21, pp. 85–95. Springer, Cham (2018)
9. Klote, J.H., Milke, J.A., Turnbull, P.G., Kashef, A., Ferreira, M.J.: Handbook of Smoke Control Engineering. American Society of Heating, Refrigerating and Air-Conditioning Engineers, Atlanta (2012)
10. Walczyk, T.: Warunki Operacyjno-Techniczne Usuwania Ciepła i Dymu z Poziomów Peronów Dworców PKP Warszawa Śródmieście i Warszawa Centralna. TEX AB, Warsaw (2003)
11. British Standards: PD 7974-6:2004. The Application of Fire Safety Engineering Principles to Fire Safety Design of Buildings. Part 6: Human Factors: Life Safety Strategies - Occupant Evacuation, Behavior and Condition. BSI, United Kingdom (2004)

12. European Parliament: Directive 2004/54/EC of the European Parliament and of the Council of 29 April 2004 on Minimum Safety Requirements for Tunnels in the Trans-European Road Network. European Parliament, Brussels (2004)
13. Baltzer, W.: Ausstattung und Betrieb von Strassentunneln: Die Neuen RABT. Strassenverkehrstechnik **61**(1), 15–22 (2017)
14. Pulsfort, M., Walz, B.: Tunnelbauverfahren, Unterirdischen Bauen, Grundbau, Bodenmechanik. Bergische Universitaet Gesamthochschule, Wuppertal (1999)
15. Nawrat, S., Schmidt-Polończyk, N., Napieraj, S.: Safety assessment of road tunnels with longitudinal ventilation, during a fire incident, utilizing numerical modelling tools. Fire Eng. **43**(3), 253–264 (2016)
16. Generalna Dyrekcja Dróg Krajowych i Autostrad. https://www.gddkia.gov.pl/a/21644/Generalny-Pomiar-Ruchu-2015-wyniki
17. Nowak, Ł., Schmidt-Polończyk, N.: Weryfikacja Możliwości Bezpiecznej Ewakuacji z Tunelu Drogowego w Warunkach Pożaru. Fire Eng. **47**(3), 96–110 (2017)
18. Hung, C.M.: Tunnel Ventilation System by (Programmable Logic Controller) PLC Control. Hong Kong Polytechnic University, Hong Kong (2008)
19. Król, M., Król, A.: Badanie Wentylacji Pożarowej w Tunelu Drogowym Laliki. In: Międzynarodowa Konferencja Ochrona Przeciwpożarowa - Zakopane Wiosna 2017, pp. 1–12. Stowarzyszenie Inżynierów i Techników Pożarnictwa, Oddział Katowice, Katowice (2017)

# Market Factors Influenced for Air and HSR Services. Case Study at the Route London - Paris

Stanisław Miecznikowski and Dariusz Tłoczyński[✉]

Faculty of Economics, University of Gdańsk, Gdańsk, Poland
stanmiecznik@gmail.com, dariusz.tloczynski@ug.edu.pl

**Abstract.** One of the main factors influencing the economic development of the region, country and the world is transport. There is competition among transport operators at the same sector, and with other market segment operators. The competition is observed among railways and air operation too. They would like to carry out, not only individual passenger's needs but also to gain higher market share and better financial results. The paper presents the functioning of two substitutable transport markets handling passenger transport at the London-Paris route and return. The aim of the research is to find out how railway and aviation operators compete in the markets to achieve market advantage. The market factors are basic purpose, the authors performed, analysed and establish two operators. For pursuing the purpose, the authors used reachable information from the subjects of the market. The main applied methods were critical in analysis of the data and the expert analysis.

**Keywords:** Competitive · Transport market · High speed railway
Air transport

## 1 Introduction

The rail and air transport markets have a big impact on the economy. In the literature, the influence of selected markets on the economies of countries, regions or cities is very often studied. The theory of economics indicates two types of functioning of the air and rail market. On the one hand, we deal with inter-industry competition, and on the other hand, rail transport plays the role of transport to the airports.

The first research aspect will be discussed in this article. It is particularly important, because despite some recourse on the railway market, there are particular connections on which rail carriers achieve a market advantage over air transport.

Both the air and rail transport market are dominated by several enterprises with significant market shares, hence the functioning of entities in a given branch takes on the characteristics of an oligopoly.

E. Macioszek and G. Sierpiński (Eds.): Directions of Development of Transport Networks and Traffic Engineering, LNNS 51, pp. 41–51, 2019.
https://doi.org/10.1007/978-3-319-98615-9_4

## 2    Market Structure in the Air and Rail Transport Industry

There are four traditional structures that form a spectrum of competition. Perfect competition lies at the competitive extreme of this spectrum and pure monopoly lies at the non-competitive extreme. Economic literature shows and defines all structures of competition [1]. In between these there are the structures of monopolistic competition and oligopoly that appear on the spectrum as illustrated in Fig. 1.

**Fig. 1.**  Structure market (Source: [1])

In air transport market and rail transport market dominate the oligopoly structure. The oligopoly is a structure where a few large sellers function and as such it lies towards the uncompetitive extreme of the spectrum of competition. It's that form of imperfect competition where there are a few firms in the market, producing either a homogeneous product or a differentiated product [2], e.g. air service and rail service. Enterprises compete with one another, and there is no collusion between the firms. D. N. Dwivedi called the oligopoly models non- collusive models [3].

In oligopoly countervailing power, Galbraith argues, prevents a large business from fully exploiting customers: "In a typical modern market of a few sellers, the active restraint is provided by competition but from the other side of the market by strong buyers … At the end of virtually every channel by which consumer goods reach the public there is, in practice, a layer of powerful buyers".

Figure 2 shows various types of direct and countervailing powers that may affect the market decisions of an undertaking providing transport services, The figure is relatively comprehensive, embracing, ideas such as countervailing power, regulations, the fear of public disquiet over excessive prices, contestability and so on as well as original competition [4].

Oligopolistic industries are typically characterized by high barriers to entry. These usually take the form of substantial capital requirements, the need for the technical and technological know-how, control of patent rights, and so forth.

In addition to few sellers in the inter-branch transport, a similar product (transport services), and high obstacles to entry, oligopolistic industries tend to share several other characteristics:

- substantial economies of scale,
- growth through merger,
- mutual dependence,
- price rigidity and non-price competition,
- the availability and cost of information,
- technical and commercial innovation,
- the numbers of buyers and sellers, and their respective power [5, 6].

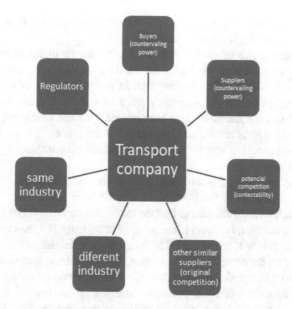

**Fig. 2.** Forces affecting a transport firm's market decisions (Source: [4])

## 3  Competitive Level at the Transport Market

There are many differences at the sharing in transport services. In its subject structure, operators put forward their services for acceptation to all segments of the market. This passenger's preferences are the basis of the offer not only which differs in the services, but also supplementary services. However, in the subjective structure the transport market there are complementary and substitutable processes. Taking into account the substitutable ones the operators compete at the different market segmentation e.g. the same sector and the other sectors.

The classical cost benefit analysis - the CBA basically estimates savings in transport cost under the assumption of perfect market. It presents the results when these transport cost savings were integrally transmitted to the others. Within such a framework, at the end of the day, this cost gains are translated into productivity gains in intangibles such as external effects or non-marketable services. However, this equivalence between transport cost saving and gains for productivity or intangibles is challenged in the presence of market imperfections [7, 8].

That is why these phenomena needs to be carefully addressed.

For competitive level the recommendation are that is necessary to analyses them, take them into account and assess the changes that the project may induce in this field. The authors insisted on the importance of the potential consequences of competitive level within the transport sector (pricing, market segmentation, frequency and level of services etc.).

Illustrative simulations of order and magnitude of pricing effects are given for the competition between air transport and high speed rail services. As regards competitive

level effects downstream of transport, no systematic correction is introduced but a
sensitivity analysis is highly recommended.

There are many terms of competitive level in economic literature but the typology
of the analysed term defined by Porter has made great influence on the opinions of
Polish and foreign authors. That can be seen in the proposition of others about their
characteristic market power [9]. Lambin's point of view is that the competitive level
term is applied to characteristics and brands that give the enterprise power against
competitors. These characteristic can be applied to the product (its basis functions) as
well as additional function tied to the product, but also the technology of production
and the form of sale at the market segment [10].

The presumption is that the competitive level has a relative character and is applied
to the competitor to achieve the best position at the market segment.

In the economics literature we can find the term competitive level by Pierścionek. His
suggestion is that competitive level is the result of knowledge (technological, marketing,
strategic) and the ability to use it [11]. The integration of the definitions by Lambin and
Pierścionek may be the most detailed interpretation of the term competitive level.

According to Tłoczyński the operators want to achieve competitive level through
competition with each other for financial assets that are owned by passenger of
transport services but they also want to gain the trust and loyalty. They take many
activities to make customers buy their goods or services [12].

To achieve competitive level and to adapt competition instruments, potential elements
of competition e.g. abilities and assets must be present. In this part the assets influence and
abilities for achievement competition competitive level are presented (Fig. 3).

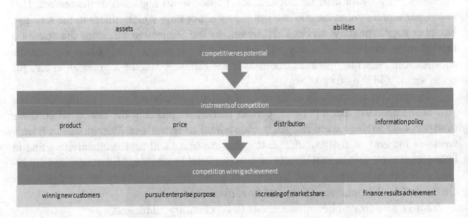

**Fig. 3.** The influence an assets and abilities for competitive level achievements

The competitive level can be analysed in different ways depending on its subject
and the point of view. The narrow point of view can be applied to: products, services,
markets or firms. The competitive level of companies and their markets is undertaken to
assess the market competition position. The air and rail transport companies establish
different instruments of action on the market [13].

## 4  Dependences of Developing Air and HSR Transport of Passengers in Europe

The development of transport depends on the growth of economy. These dependence is more important in the case of freight transport than passenger transport.

During twenty one years, we can see almost constant growth of the EU passenger transport measured in RPKs, without any influence the GDP growth in 2007–2009 years. The freight transport measured FTKs, otherwise dramatically slowed down in the period of the EU recession. After recession FTKs freight transport grew faster than RPKs passenger transport in the year 2011, then in 2012 it slowed down in comparison to the passenger year rate growth. From then until 2015 the transport freight growth was slower than the growth of passenger transport in RPKs. As a result both means of transport equalized their growth in 2015.

In 2014 and 2015, the rate growth of passenger transport was 2.6 per cent, at the same time FTKs freight transport only 1.2 per cent. In the period 2013–2015, the rate growth of passenger transport was two times bigger than freight transport [14].

Let us see what kind of passenger transport sectors in the EU gained the biggest market share in the analysed period (Fig. 4).

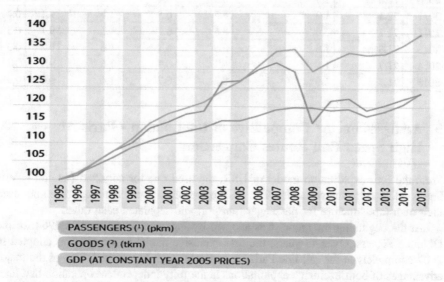

PASSENGERS (¹) (pkm)

GOODS (²) (tkm)

GDP (AT CONSTANT YEAR 2005 PRICES)

Notes: (¹)  Passenger cars, powered two-wheelers, buses & coaches, tram & metro, railways, intra-EU air, intra-EU sea.
(²)  Road, rail, inland waterways, oil pipelines, intra-EU air, intra-EU sea.

GDP: at constant year 2005 prices and exchange rates.

**Fig. 4.** Transport growth to GDP's in the years 1995–2015, Year 1995 = 100 (Source: [14])

As we see in Table 1, the first place got the road transport (passenger cars, bus & coach, P2W) but its share declined gradually, but it did not lose its biggest market's share. However, the air transport market share was growing up while rail transport was fixed. It's possible to make judgement that operators of air transport achieved more competitive level than railways to get more passengers in the EU market. Can it be proved at the selected route?

**Table 1.** Modal split at the passenger's EU market in years 1995–2015 (per cent) (Source: [14]).

| Year | Passenger cars | P2W (Powered two wheeler) | Bus & Coach | Railway | Tram & Metro | Air | Sea |
|------|------|------|------|------|------|------|------|
| 1995 | 73.2 | 2.1 | 9.6 | 6.6 | 1.4 | 6.5 | 0.6 |
| 2000 | 72.9 | 1.8 | 9.3 | 6.3 | 1.4 | 7.8 | 0.5 |
| 2005 | 72.7 | 1.9 | 8.8 | 6.1 | 1.4 | 8.5 | 0.5 |
| 2006 | 72.5 | 1.9 | 8.7 | 6.2 | 1.4 | 8.8 | 0.4 |
| 2007 | 72.3 | 1.8 | 8.8 | 6.2 | 1.4 | 9.0 | 0.4 |
| 2008 | 72.0 | 1.9 | 8.9 | 6.4 | 1.5 | 8.8 | 0.5 |
| 2009 | 73.2 | 1.8 | 8.6 | 6.3 | 1.5 | 8.2 | 0.4 |
| 2010 | 72.8 | 1.9 | 8.5 | 6.4 | 1.5 | 8.5 | 0.4 |
| 2011 | 72.1 | 1.9 | 8.5 | 6.5 | 1.5 | 9.1 | 0.4 |
| 2012 | 71.7 | 2.0 | 8.6 | 6.7 | 1.6 | 9.1 | 0.3 |
| 2013 | 71.8 | 1.9 | 8.5 | 6.7 | 1.6 | 9.2 | 0.3 |
| 2014 | 71.7 | 1.9 | 8.3 | 6.7 | 1.6 | 9.5 | 0.3 |
| 2015 | 71.5 | 1.9 | 8.2 | 6.7 | 1.6 | 9.8 | 0.3 |

## 5    Analyses Factor Competitive at the London - Paris Air - High Speed Rail Routes Connection

From the four exploitation model of HSR, the used one for admitted London - Paris line is adopted exclusive model. The HSR operators and conventional operators used their own infrastructure for passenger transport and separated each other.

At the beginning the model was adopted by Japanese Shinkansen in 1964 for the Olympic Players. Over 43 years, the first exclusive model in Europe was adopted in 2007 completely at the rail line Paris - London through Eurotunnel. One of the major advantages of both exclusive exploitations is the fully independent operation that later provides assets possibilities for privatization.

The UK government sold its 40 per cent in train cross channel operator Eurostar to the Anglo - Canadian consortium for 730 mln pounds. The deal went through and they owned respectively 30 per cent and 10 per cent of Eurostar.

Eurostar began service in 1994 as a partner between three state railways: France (SNCF) - 55% stake, Belgium (SNCB) - 5% stake and British Rail 40% stake subsequently Long Continental Railway. Since then Eurostar has carried more than 160 million passengers and more than 10 million in 2015 alone.

There are three types of supply service models of air and rail transport for passengers far distance demand mobility growth (Fig. 5). It has been chosen multimodal transport supply service at the admitted routes. There are differences between two modes of transport at the beginning and end points of travel. The HSR's has only one in London St. Pancras and Grand Garde in Paris. The Paris and London have many airports.

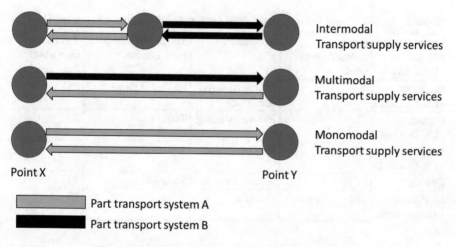

**Fig. 5.** Transport supply service models (Source: [15])

Every half an hour there are two train departures from the London St. Pancras HSR railway station from 5 a.m. to 10 p.m. to Paris. It's useful for increasing passenger's accessibility for the Eurostar services.

The London agglomeration has 5 airports: Heathrow (78 million pax in 2017), Gatwick (46 million pax), Luton (16 million pax), Stansted (25 million pax) and City (5 million pax), serving a total of approximately 171 million pax. Analyzing the profile of carriers operating in the ports, it can be noticed that Heathrow and City are the ports serving traditional carriers, while at Gatwick, Luton and Stansted airports dominates the low-cost traffic. Of all the airports mentioned, with the exception of Stansted, carriers offer connections to Paris. The main operators are Air France, British Airways and EasyJet. They carried almost 2 million passengers in 2017.

Almost 20 flights a day are offered by carriers from four English airports, most from Heathrow Airport. It should be noted that from Stansted Airport Rynaiair does not offer any connection to Paris.

When analyzing the data in Table 2, two types of offered transport services should be distinguished. Business transfers that AF and BA carriers offer from Heathrow Airport and City, as well as commercial flights offered by LCC from Luton Airport and Gatwick Airport. The supply of business transport is 116 connections during the week. A characteristic feature of these connections is the short journey time to the airport. This short time of travel to the airport at high frequency along with the flight time is designed to create a competitive offer in relation to rail transport. At the same time, the price for transporting passengers using traditional carriers is much higher than for low-cost carriers. The offer of transport from the ports of Gatwick and Luton is directed to passengers traveling on a tourist trip. The price is also adjusted to this segment. The average prices for passenger transport on the London-Paris route are presented in Table 3.

**Table 2.** Supply air and rail connection from London to Paris (Source: based on: [16–21]).

| Route: London Airports - Paris Airports | Direct connections per week | Air carriers | The fastest transport from London | The fastest time travel from London |
|---|---|---|---|---|
| Heathrow - Ch. De Gaulle | 99 | British Airways, Air France | Heathrow Express | 15 |
| Gatwick - Ch. De Gaulle | 32 | EasyJet, Vueling | Train | 50 |
| Luton - Ch. De Gaulle | 20 | EasyJet | Train | 50 |
| City-Orly | 17 | British Airways | Train DLR | 20 |
| St. Pancras - Grand Garde | 102 | Eurostar | Tube | 10 |

**Table 3.** Average prices for travel and travel time in London - Paris [March 2018] (Source: based on: [16–21]).

| | LCC | Economy class | Business class | Travel time |
|---|---|---|---|---|
| Air transport | 80 GBP | 220 GBP | 340 GBP | 75–90 min |
| HSR | 171 GBP (standard) | 247 GBP (standard premier) | 276 EUR | 140 min |

Analyzing the competitive level of airplane and rail carriers operating on the London-Paris route not only travel time of a particular mean of transport must be taken into consideration, but also time of travel in the relation city - airport and airport - city, and other factor different services on board (air and HSR).

Due to the specificity of London and Paris, passengers traveling for business purposes are the dominant segment of travellers. The offer of business transport is diversified very well in the case of rail and air connections. The extensive tariff systems or the number of available HSR and Air transport connections indicate that such connections are profitable, bringing large incomes to operators. Despite the large offer for travellers from the business sector, also passengers travelling for tourism use the possibility of moving between London and Paris by air or rail. The choice of a particular means of transport depends on individual travel expectations in terms of: departure and arrival times (arrival and departure), arrival time, level of security, speed of reaching the destination and to the starting point of travel, and thus the total travel time.

Passengers on a business trip expect to have a short journey time, direct access, availability, convenient departure and arrival times. The price does not affect the decision about the choice of transport. As a result of comparison and evaluation of railway and air connections, the most important factors of the competitive level of carriers operating on the London-Paris route should be pointed out (Tables 4 and 5).

**Table 4.** Services offer by air carriers and HSR (Source: based on: [16–21]).

| Air transport | LCC | Economy class | Business class |
|---|---|---|---|
| | • All our standard flight prices are shown for one adult travelling, one way and include all taxes, fees (including our £15 administration fee), charges and cabin bag<br>• Hold luggage is not included and can be added for an additional fee | • You'll have an allocated seat when you check in and it includes our generous hand baggage allowance<br>• With Basic fares, you have the option of paying to add a checked bag, choose your seat, or other extras facilities | • Perfect combination of efficiency, convenience and comfort<br>• Access to comfortable departures lounges<br>• Be one of the first on board with priority boarding<br>• More personal space on board to work or relax<br>• Complimentary food and drink service<br>• Dedicated check-in desks and priority boarding<br>• Larger baggage allowance than economy class<br>• Collect more loyalty programms |
| HSR | Standard | Premier standard | Business premier |
| | • 2 pieces of luggage + 1 hand luggage<br>• Change journey with a fee<br>• Arrive at station 45–60 min before departure<br>• Buy snacks, drinks and meals on board | • 2 pieces of luggage + 1 hand luggage<br>• Change journey with a fee<br>• Arrive at station 45–60 min before departure<br>• Light meal and drinks served at your seat<br>• Spacious seats | • 3 pieces of luggage + 1 hand luggage<br>• No change fees and free cancellation<br>• Ticket gate closes 10 min before departure<br>• Hot meals designed by Raymond Blanc, served with champagne<br>• Spacious seats<br>• Exclusive lounge<br>• Taxi booking service |

**Table 5.** Air and rail services assessment (Source: based on: [16–21]).

| Factor | Air transport | HSR |
| --- | --- | --- |
| Direct | Yes | Yes |
| Travel time | Quickly | Good |
| Available airports/station | Heathrow and City - very good, Luton and Gatwick - middle | In city |
| Time departures/arrives | Heathrow and City - very good, Luton and Gatwick - weak | Very good |
| Fare | Differences for class | Differences for class |
| Luggage | Differences for class | Yes |
| Facilities | Differences for class | Differences for class |

Basing on the research and analyzes of the competitiveness level, it is not possible to state clearly which branch is more competitive or which has a higher level of competitiveness. The choice of the carrier and at the same time the transport branch results from the entire individualized carrier's offer, at a specific time and individual expectations of the buyers.

# 6  Summary

As a result of the conducted research on the functioning of the oligopolistic passenger transport market on the London - Paris route, the following conclusions should be drawn:

- in the structure of the passenger transport market in question, there is the largest competition in transport in the business segment. For this segment air carriers and rail carriers offer the largest supply of transport services,
- it should be pointed out that transport services in the business segment, both rail and air, are offered from the center of London. In the case of flights from Heathrow and City, the travel time to the airport is acceptable to the business traveller,
- despite the fact that the offer addressed to the business segment is characterized by high prices, it is fully acceptable by passengers. The business segment very often decides whether to start a journey and choose a mode of transport a few hours before the trip. Hence, it adopts a high market price for the services provided. In return, passengers receive a high standard of services,
- the level of competitiveness of HSR carriers and air carriers is also influenced by the availability of the service, convenient times of the journey and times of reaching the destination and comfort on board the means of transport,
- greater attractiveness in the economy segment is characterized by trips made by rail, above all to a higher frequency, proximity to the city center and a larger travel package compared to the offer of LCC carriers. However, the price for transporting passengers by rail is much greater than the service provided by lost cost carriers.

# References

1. Mallard, G., Glaister, S.: Transport Economics. Theory, Application and Policy. Palgrave MacMillan, New York (2010)
2. Mandal, R.K.: Microeconomisc Theory. Atlantic Publishers & Distributors, New Delhi (2007)
3. Dwivedi, D.N.: Microeconomisc: Theory and Applications. Dorling Kinderslay, London (2008)
4. Button, K.: Transport Economics. Edward Elgar, Northampton (2010)
5. Holloway, S.: Straight and Level: Practical Airline Economics. Ashgate, Burlington (2003)
6. Wensveen, J.G.: Air Transportation. A Management Perspective. Ashgate, Burlington (2007)
7. International Transport Forum: Improving the Practice of Cost Benefit Analysis in Transport. http://dx.doi.org/10.1787/5kghzxq2q546-en
8. De Palma, A., Lindsey, R., Quinet, E., Vickerman, R.: A Handbook of Transport Economics. Edward Elgar Publishing Limited, Northamptom (2011)
9. Porter, M.E.: Competitive Advantage. Creating and Sustaining Superior Performance. The Free Press, New York (1998)
10. Lambin, J.J., Schuiling, I.: Market-Driven Management: Strategic and Operational Marketing. Palgrave Macmillan, London (2012)
11. Pierścionek, Z.: Strategie Konkurencji i Rozwoju Przedsiębiorstwa. PWN, Warsaw (2003)
12. Tłoczyński, D.: Konkurencja Na Polskim Rynku Usług Transportu Lotniczego. Gdańsk University Publisher, Gdańsk (2016)
13. Rucińska, D.: Rynek Usług Transportowych w Polsce. PWE, Warsaw (2015)
14. Directorate - General for Mobility and Transport (European Commission): EU Transport in Figures. Statistical Pocketbook 2017. Publications Office of the European Union, Luxembourg (2017)
15. Souter-Servaes, T.: Unkomplizierte Komplizen - Neue Wege der Luft - Schiene-Kooperation. Internationales Verkehrswesen, Hamburg (2007)
16. Heathrow Airport. http://www.heatrow.com
17. Gatwick Airport. http://www.gatwickairport.com
18. Luton Airport. http://www.london-luton.co.uk
19. Stansted Airport. http://www.stanstedairport.com
20. London City Airport. http://www.londoncityairport.com
21. Eurostar. http://www.eurostar.com

# Directions of Improvement in Transport Systems

# The Condition of EV Infrastructure in the World - Analysis for Years 2005–2016

Ewelina Sendek-Matysiak[✉]

Faculty of Mechatronics and Machine Design, Kielce University of Technology,
Kielce, Poland
esendek@tu.kielce.pl

**Abstract.** Many prominent environment protection activists, political decision-makers and businessmen think that widespread global use of electric vehicles (EV) is necessary to mitigate many problems connected with environment protection, energy security and sustainable development [1, 2]. In spite of intense promotional actions carried out by numerous governments and vehicle manufacturers, global diffusion of EV is generally slow [3]. One of the main barriers in the development of market for these cars includes limited transport infrastructure dedicated to vehicles of this type - that is public charging stations for electric vehicles. This issue is important in so far as it concerns their quantity, parameters, and availability. In this article, the author will discuss the condition of EV infrastructure in the world as regards the number of charging stations, as well as relevant preconditions to increase their number.

**Keywords:** Electromobility · Sustainable development policy
Electric vehicle · Charging station

## 1 Introduction

The environmental impact of transport, being a very important part of sustainable development strategy, is listed among the main green transport objectives of the European Union.

Currently, the transport sector is one of the largest contributors to carbon emissions, and despite the fact that transport related emission levels have been decreasing since 2007, they have still not reached the level of 1990.

This makes it necessary for transport systems to shift from fossil fuels to alternative energy sources less dependent on fossil fuel energy. The key to this change is electric mobility, in particular the introduction of electric vehicles (EV). Fully or partially electrically powered vehicles reduce both the dependence on fossil fuels and green-house gas emissions, thus noticeably contributing to the green growth which, according to the White Paper of 2011, The "Roadmap to a Single European Transport Area - Towards a competitive and resource efficient transport system" by 2050 will have reduced $CO_2$ emissions by 60% [4].

However, although these cars offer many advantages including lower operation costs, better acceleration, less maintenance requirements, lower noise emission level than in case of cars with internal combustion engines, and first of all zero pollution

© Springer Nature Switzerland AG 2019
E. Macioszek and G. Sierpiński (Eds.): Directions of Development of Transport
Networks and Traffic Engineering, LNNS 51, pp. 55–65, 2019.
https://doi.org/10.1007/978-3-319-98615-9_5

emission at the place they are operated, which is characteristic for them, the share of EV in the automotive market still remains slight.

In 2016, total number of new light-weight electric cars registered in the world exceeded 753 000 and although the increase was 38% compared to 2015, we observed slowdown in the growth rate compared to previous years (Fig. 1). That was so because in 2016, first time after 2010, the year-to-year increase in passenger EV sale dropped under 50%.

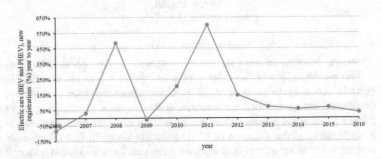

**Fig. 1.** The year-to-year sale of electric cars (Battery Electric Vehicle (BEV) and Plug-in Hybrid Electric Vehicle (PHEV)) in years 2006–2016 (Source: own study on the basis of [5])

Tables 1 and 2 contain data on new registrations and the share of electric cars in the market in years 2005–2016, respectively.

Definitely, in 2016 the most of electric cars were registered in China (336 000), and this exceeded more than twice their number in the United States (159 000). At the same time, in Europe this parameter fluctuated around 215 000.

**Table 1.** Electric cars (battery electric and plug-in hybrid), new registrations by country, 2005–2016, (thousands) (Source: own study on the basis of [5, 6]).

| Country | 2005 | 2006 | 2007 | 2008 | 2009 | 2010 | 2011 | 2012 | 2013 | 2014 | 2015 | 2016 |
|---|---|---|---|---|---|---|---|---|---|---|---|---|
| Canada | | | | | | | 0.52 | 2.02 | 3.12 | 5.07 | 6.96 | 11.58 |
| China | | | | | 0.48 | 1.43 | 5.07 | 9.9 | 15.34 | 73.17 | 207.3 | 336 |
| France | 0.01 | | | | 0.1 | 0.19 | 2.73 | 6.26 | 9.62 | 12.64 | 22.95 | 29.51 |
| Germany | 0.02 | | | 0.07 | 0.02 | 0.14 | 1.65 | 3.37 | 6.93 | 12.74 | 23.19 | 24.61 |
| India | | | | 0.37 | 0.15 | 0.35 | 0.45 | 1.43 | 0.19 | 0.41 | 1 | 0.45 |
| Japan | | | | | 1.08 | 2.44 | 12.62 | 24.44 | 28.88 | 32.29 | 24.65 | 24.85 |
| Korea | | | | | | 0.06 | 0.27 | 0.51 | 0.6 | 1.31 | 3.19 | 5.26 |
| Netherlands | | | | 0.01 | 0.03 | 0.12 | 0.88 | 5.12 | 22.42 | 15.09 | 43.77 | 24.48 |
| Norway | | | 0.01 | 0.24 | 0.15 | 0.39 | 1.84 | 4.51 | 8.52 | 19.76 | 35.61 | 50.18 |
| Poland | | | | | | | 0.035 | 0.03 | 0.04 | 0.15 | 0.26 | 0.27 |
| Sweden | | | | | | | 0.18 | 0.93 | 1.55 | 4.67 | 8.59 | 13.42 |
| United Kingdom | 0.22 | 0.32 | 0.45 | 0.22 | 0.18 | 0.28 | 1.22 | 2.69 | 3.75 | 14.74 | 29.34 | 37.91 |
| United States | 1.12 | | | 1.47 | | 1.19 | 17.73 | 53 | 97 | 118.7 | 113.8 | 159.6 |
| Others | 0.53 | | | 0.08 | 0.03 | 0.18 | 2.39 | 3.61 | 6.01 | 12.62 | 26.35 | 35.04 |
| Total | 1.9 | 0.32 | 0.46 | 2.46 | 2.22 | 6.77 | 47.59 | 118.1 | 203.7 | 323.4 | 547.1 | 753.2 |

Besides purchase cost and average driving range, the essential barriers for the growth of electric car market include limited transport infrastructure dedicated for these vehicles, that's public charging points for electric vehicles. This issue is so important because it concerns their number, parameters, and their availability as well.

Difficulties of this sort may be encountered not only in Poland, but all over the world as well.

**Table 2.** Electric cars (battery electric and plug-in hybrid), market share by country, 2005–2016 (expressed as percentage) (Source: [7]).

| | 2005 | 2006 | 2007 | 2008 | 2009 | 2010 | 2011 | 2012 | 2013 | 2014 | 2015 | 2016 |
|---|---|---|---|---|---|---|---|---|---|---|---|---|
| Canada | | | | | | | | 0.15 | 0.20 | 0.29 | 0.39 | 0.59 |
| China | | | | | | 0.01 | 0.04 | 0.06 | 0.09 | 0.38 | 0.99 | 1.37 |
| France | | | | | | 0.01 | 0.13 | 0.34 | 0.55 | 0.72 | 1.22 | 1.46 |
| Germany | | | | | | 0.00 | 0.05 | 0.11 | 0.23 | 0.42 | 0.72 | 0.73 |
| India | | | | 0.02 | 0.01 | 0.01 | 0.02 | 0.05 | 0.01 | 0.02 | 0.04 | 0.02 |
| Japan | | | | | 0.03 | 0.06 | 0.35 | 0.53 | 0.63 | 0.68 | 0.58 | 0.59 |
| Korea | | | | | | | 0.02 | 0.04 | 0.05 | 0.09 | 0.21 | 0.34 |
| Netherlands | | | | | | 0.01 | 0.02 | 0.16 | 1.02 | 5.38 | 3.89 | 9.74 | 6.39 |
| Norway | | | 0.01 | 0.22 | 0.15 | 0.31 | 1.33 | 3.27 | 6.00 | 13.7 | 23.6 | 28.7 |
| Poland | | | | | | | 0.01 | 0.01 | 0.05 | 0.07 | 0.06 | 0.19 |
| Sweden | | | | | | 0.00 | 0.05 | 0.31 | 0.53 | 1.44 | 2.37 | 3.41 |
| United Kingdom | 0.01 | 0.01 | 0.02 | 0.01 | 0.01 | 0.01 | 0.06 | 0.13 | 0.17 | 0.60 | 1.11 | 1.41 |
| United States | 0.01 | | | 0.01 | | 0.01 | 0.17 | 0.44 | 0.75 | 0.74 | 0.67 | 0.91 |
| Others | 0.01 | 0.00 | 0.00 | 0.00 | 0.00 | 0.00 | 0.03 | 0.05 | 0.05 | 0.14 | 0.32 | 0.33 |
| Total | 0.00 | 0.00 | 0.00 | 0.01 | 0.01 | 0.01 | 0.10 | 0.23 | 0.38 | 0.54 | 0.85 | 1.10 |

Limited number and availability of public charging points (EVSE - electric vehicle supply equipment) or EV chargers (EVC) induce fears among drivers, which as a consequence affects their decisions regarding selection of car with certain drive type. It is therefore vital to achieve sufficient charging infrastructure.

## 2   EV Charger Characteristics

Most generally, electric recharging points can be divided depending on the presence or lack of wires for electric energy transmission between vehicles and chargers. This classification allows distinguishing conductive plug-in charging stations and wireless charging stations [8]. This second group is also called inductive, as in their case electric energy is transmitted by magnetic fields through inductive coupling among two or more coils, which work as a transformer with wide air-gap [9].

Electric vehicle charging, where plug-in charging stations are used, is defined in a number of standards developed by international standardisation bodies including the

International Organization for Standardisation (ISO), the International Electrotechnical Commission (IEC), Society of Automotive Engineers (SAE) and Standardization Administration of China (SAC) in cooperation with CHAdeMO and Charin associations.

Standards specified in Fig. 2 are defined on the basis of parameters characteristic for a charging point, there are:

- level, describing the power output of an EVSE outlet,
- type, referring to the socket and connector being used for charging,
- mode, which describes the communication protocol between the vehicle and the charger.

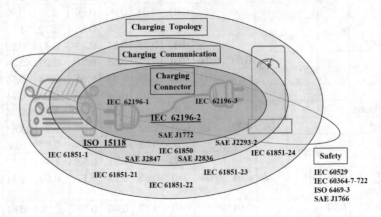

**Fig. 2.** Standards for electric car charging operation (Source: own study on the basis of [7])

Figure 3 lists possible levels (power output) and types (socket and connector) of EVSE used in China, Europe, Japan and the United States.

| Classification in use here | Level | Current | Power | Type | | | |
|---|---|---|---|---|---|---|---|
| | | | | China | Europe | Japan | North America |
| Slow chargers | Level 1 | AC | ≤3.7 kW | Devices installed in private households, the primary purpose of which is not | | | SAE J1772 Type 1 |
| | Level 2 | AC | > 3.7 kW and ≤ 22 kW | GB/T 20234 AC | IEC 62196 Type 2 | SAE J1772 Type 1 | SAE J1772 Type 1 |
| | Level 2 | AC | ≤ 22 kW | Tesla connector | | | |
| Fast chargers | Level 3 | AC, triphase | > 22 kW and 43.5 kW | | IEC 62196 Type 2 | SAE J3068 (under development) | |
| | Level 3 | DC | Currently < 200 kW | GB/T 20234 DC | CCS Combo 2 Connector (IEC 62196 Type 2 & DC) | CHAdeMO | CCS Combo 1 Connector (SAE J1772 Type 1 & DC) |
| | Level 3 | DC | Currently < 150 kW | Tesla and CHAdeMO connectors | | | |

**Fig. 3.** Overview of the level (power output) and type (socket and connector) of EVSE used in China, Europe, Japan, and the United States (Source: own study on the basis of [10])

Available power level in charging point has significant impact on electric vehicle battery charging time. In case of level 1 the charger is an internal component of a car. Alternating current is transmitted from the distributor via standard socket, 1-phase, 230 V. Power available from converter is being limited to 2 kW, which results in battery charging time ranging from 11 to 14 h, depending on its capacity. In case of level 2 the charger also fits inside a car. The vehicle is being charged with alternating current, either 1- or 3-phase. The system power may reach up to 22 kW, which allows reducing charging time to 2–3 h. As regards level 3, here the charger is located outside the EV. The vehicle battery terminals are connected to a special connector installed in the car. Battery is charged with direct current. The system power reaches even 50 kW. This method allows charging up to 80% of the battery capacity within only 15–30 min, and full charge takes up to 1 h [11].

# 3  Own Study

Total number of the EVSE charging stations is permanently growing with the number of light-weight electric cars, and in 2016 it reached 322 265, among which 212 394 were slow charging points (>3.7 kW and ≤ 22 kW), and 109 871 - fast charging points (other) (Figs. 4, 5, 6, 7, 8 and 9).

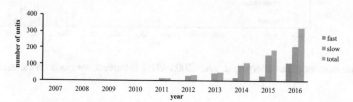

**Fig. 4.** The number of generally available charging points in the world in years 2007–2016 (in thousands) (Source: [5])

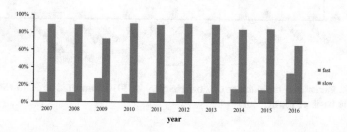

**Fig. 5.** The share of slow and rapid charging points in the world in years 2007–2016 (Source: own study on the basis of [5])

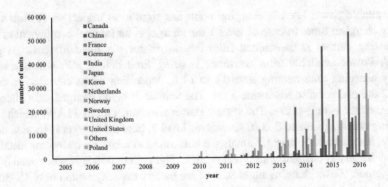

**Fig. 6.** The number of slow chargers in years 2005–2016 (Source: own study on the basis of [5, 6])

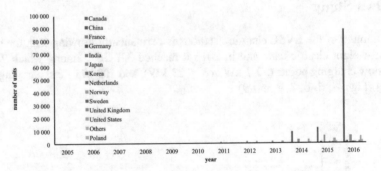

**Fig. 7.** The number of fast chargers in years 2005–2016 (Source: own study on the basis of [5, 6])

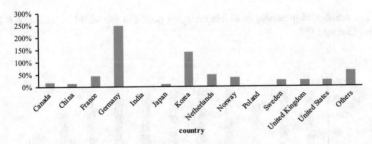

**Fig. 8.** The increase in the number of slow chargers in 2016 compared to 2015 (Source: own study on the basis of [5, 6])

First of all, the increase in the number of public recharging points in 2016 by 72% compared to 2015 results from sudden increase in the number of fast chargers (292% compared to previous year). It should be pointed out that for the most part this result was achieved after introduction of fast chargers in China, where their number increased seven times (631%) compared to previous year. Even these are omitted, the number of public charging points in the world grew by 41% compared to 2015 (42% - slow and

34% - fast). Greater increase in the number of public charging stations in 2016 (72%) compared to the increase in the number of electric vehicles EV (60%) is most desirable. However, still the number of electric cars per one EV charging point is high. This indicates that majority of drivers still use their home recharging units, which considerably extends their vehicle charging time, which in consequence may deter people from using these vehicles. In 2016, there were six electric cars in the world per one public charging point, nine EV per one slow charging point, and 18 eV per fast charging point (Figs. 10, 11 and 12). The most light-weight electric cars per one EVSE are reported in Norway and the USA, which doubtlessly is the outcome of highest share of electric cars in the automotive markets of these countries.

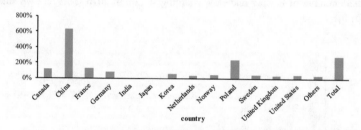

**Fig. 9.** The increase in the number of fast chargers in 2016 compared to 2015 (Source: own study on the basis of [5, 6])

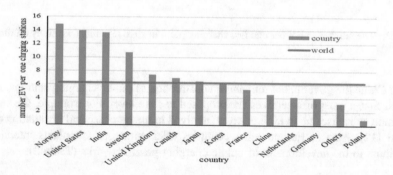

**Fig. 10.** The number of EV per one charging station in 2016 (Source: own study on the basis of [5, 6])

In order to increase the number of public charging points, instruments intended to encourage local authorities, companies and physical persons to install the EVSE have been developed and introduced in many countries, and they are continuously implemented there.

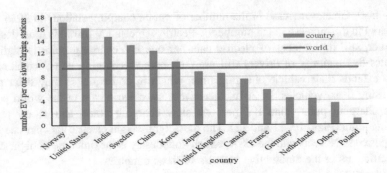

**Fig. 11.** The number of EV per one slow charging station in 2016 (Source: own study on the basis of [5, 6])

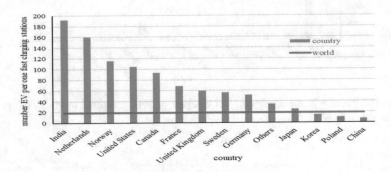

**Fig. 12.** The number of EV per one fast charging point in 2016 (Source: own study on the basis of [5, 6])

In China the government refinances construction of public charging stations. In the United States most of incentives are proposed at state level. For example, the state of Colorado subsidises up to 80% of costs involved in the purchase and installation of the EVSE [12]. Also in Europe some countries offer numerous stimulants intended to contribute to the development of public charging point networks (Table 3).

**Table 3.** Infrastructure incentives in the EU countries (Source: own study on the basis of [6]).

| Countries | Infrastructure incentives |
|---|---|
| Denmark | Tax rebate on installation of EV home chargers of up to 18.000 DKK ($2.646) <br> Connection charge reduced by 50% for public charging stations |
| Iceland | Charging Stations incentives |
| Ireland | Free installation of domestic chargers up to 2 000 |
| Italy | Credit for charging infrastructure in non-residential buildings over 500 m$^2$ |

*(continued)*

**Table 3.** (*continued*)

| Countries | Infrastructure incentives |
|---|---|
| Malta | Electric Car owners can charge their vehicles at home using residence electricity subsidies<br>Grant of €2.000 to assist companies to buy charging points - up to five charging points per company, that is the total of €10.000 grant |
| Norway | Public funding for fast charging stations every 50 km on main roads |
| Romania | A refund of maximum 2.500 euro for Stations < 22 kW and 30.000 euro for Stations > 22 kW<br>You can apply by Oct. 15 2016 and eligible are cities over 50 000 citizens |
| Spain | Subsidies for private and public charging points |
| United Kingdom | £500 incentive for installing a dedicated home charging station<br>Up to 75% (capped at £7500) towards the cost of installing an on-street residential charge point in areas without off-street parking |

# 4  Conclusions

Results of the analysis performed in this article indicate that the number of public charging points worldwide was growing steadily between 2007–2016, reaching an average annual growth of 212%. Doubtlessly, their considerable number would compensate occupancy time of such station by an electric car being charged. Table 4 specifies EV charging duration depending on charger type.

Increase in the number of EVSE will allow more cars that may be charged per hour. Currently, fast direct current charger, for which charging plug utilization rate is 50%, will be used for 12 h in a day to charge one Nissan Leaf for 31 min. This implies that in one day this unit can be used to charge more or less 23 Nissans Leaf. In case of level 1 and 2 chargers the capacity would be even smaller due to longer charge time.

Moreover, extended network of charging points would increase chances to find a charger, which would be compatible with a given car. The simplest solution of this problem is to use the multi-protocol fast charging stations; however, due to considerable costs their number is and will be limited. According to [14], the total number of EV chargers in the world will reach 12.7 million by 2020, growing at the rate of 2–3 million per year. Due to high costs of DC Fast Charging, which may range from USD 8.500 to 50.000 [15], their share will be slight, and they will be located on the outskirts of cities or by motorways.

There are alternative methods allowing EV charging (e.g. solar cells as additional charging sources, possibility to charge a vehicle while it is running (DEVC)), however they are still under tests and thus it is necessary to promote already known and proven solutions, while implementing the newer ones as they occur.

**Table 4.** Charging times for currently available EVs, based on type of charger. (Source: [13]).

| Make and model | Total range (Miles) | Time to charge [h] | | |
|---|---|---|---|---|
| | | Level 1 charger | Level 2 charger | DC charging |
| Mitsubishi i-Miev | 62 | 12.4 | 3.1 | 0.29 |
| Smart fortwo electric drive | 68 | 13.6 | 3.4 | 0.32 |
| Ford Focus Electric | 76 | 15.2 | 3.8 | 0.36 |
| BMW-i3 BEV | 81 | 16.2 | 4.05 | 0.38 |
| Chevrolet Spark EV | 82 | 16.4 | 4.1 | 0.39 |
| Volkswagen e-Golf | 83 | 16.6 | 4.15 | 0.39 |
| Fiat 500e | 84 | 16.8 | 4.2 | 0.4 |
| Mercedes-Benz B250e | 87 | 17.4 | 4.35 | 0.41 |
| Kia Soul Electric | 93 | 18.6 | 4.65 | 0.44 |
| Nissan Leaf, 24 kW-h | 84 | 16.8 | 4.2 | 0.4 |
| Nissan Leaf, 30 kW-h | 107 | 21.4 | 5.35 | 0.5 |
| Tesla Model S, 70 kW-h | 234 | 46.8 | 11.7 | 1.11 |
| Tesla Model S, 85 kW-h | 265 | 53 | 13.25 | 1.26 |
| Tesla Model S, AWD 70D | 240 | 48 | 12 | 1.14 |
| AWD 85D | 270 | 54 | 13.5 | 1.28 |
| AWD 90D | 270 | 54 | 13.5 | 1.28 |
| AWD P85D | 253 | 50.6 | 12.65 | 1.20 |
| AWD P90D | 253 | 50.6 | 12.65 | 1.20 |
| Tesla Model X, AWD 90D | 257 | 51.4 | 12.85 | 1.22 |
| AWD P90D | 250 | 50 | 12.5 | 1.19 |

# References

1. Boren, S., Nurhadi, L., Ny, H., Robert, K.H., Broman, G., Trygg, L.: A strategic approach to sustainable transport system development - part 2: the case of a vision for electric vehicle systems in Southeast Sweden. J. Clean. Prod. **140**, 62–71 (2017)
2. Mock, P., Yang, Z.: Driving Electrification. A Global Comparison of Fiscal Incentive Policy for Electric Vehicles. International Council on Clean Transportation, Washington (2014)
3. Olson, E.L.: The financial and environmental costs and benefits for norwegian electric car subsidies: are they good public policy? Int. J. Technol. Policy Manag. **15**(3), 277–296 (2015)
4. European Commission: White Paper. Roadmap to a Single European Transport Area - Towards a Competitive and Resource Efficient Transport System. European Commission, Brussels (2011)
5. International Energy Agency. www.iea.org/
6. European Alternative Fuels Observatory. www.eafo.eu/

7. Global EV Outlook 2017. Two Million and Counting. https://www.iea.org/publications/ freepublications/publication/GlobalEVOutlook2017.pdf
8. Rubino, L., Capasso, C., Veneri, O.: Review on plug-in electric vehicle charging architectures integrated with distributed energy sources for sustainable mobility. Appl. Energy **207**, 438–464 (2017)
9. Zicheng, B., Song, L., De Kleine, R., Chunting, ChM, Keoleian, G.A.: Plug-in vs. wireless charging: life cycle energy and greenhouse gas emissions for an electric bus system. Appl. Energy **146**, 11–19 (2015)
10. Soylu, S.: Electric Vehicles. The Benefits and Barriers. InTech Open Limited, London (2011)
11. Zajkowski, K., Seroka, K.: Review on available methods used to charge batteries in electric vehicles. Buses: Technol. Oper. Transp. Syst. **7–8**, 483–486 (2017)
12. Hodge, C.: Personal Communication of the Authors with Cabell Hodge. National Renewable Energy Laboratory of the United States, Washington (2017)
13. Adderly, S.A., Manukian, D., Sullivan, T.D., Son, M.: Electric vehicles and natural disaster policy implications. Energy Policy **112**, 437–448 (2017)
14. IHS Automotive. https://ihsmarkit.com/industry/automotive.html
15. Lam, A., Leung, Y.W., Chu, X.: Electric vehicle charging station placement: formulation, complexity and solutions. IEEE Trans. Smart Grid **5**(6), 2846–2856 (2014)

# Changes in Values of Traffic Volume - Case Study Based on General Traffic Measurements in Opolskie Voivodeship (Poland)

Elżbieta Macioszek[✉]

Faculty of Transport, Silesian University of Technology, Katowice, Poland
elzbieta.macioszek@polsl.pl

**Abstract.** Traffic measurements and surveys are indispensable in road traffic engineering. Results delivered by such studies are used in numerous detailed follow-up analyses, but also to make planning decisions and to design transport infrastructure as well as traffic control systems. This article addresses outcomes of a detailed analysis of results of the General Traffic Measurements conducted over the years 2000–2015. The analysis has been performed with reference to results pertaining to Opolskie Voivodeship.

**Keywords:** Traffic measurements · Annual average daily traffic
Road traffic engineering · Transport

## 1 Introduction

In contemporary times, traffic measurements and surveys are indispensable elements of road traffic engineering. Results delivered by such measurements are used in numerous analyses, but also to make planning decisions, to design transport infrastructure and traffic control systems, to plan sustainable mobility with regard to eco-friendly forms of transport as well as for purposes of many other detailed analyses (e.g.: [1–17]), including studies on the introduction of new kind of vehicles such as electric buses [18–20]. Furthermore, the main goal of all kinds of road traffic surveys is to obtain information on the nature of traffic and the regularities that govern it. Measurements are also intended to deliver results required for various kinds of analyses or expert opinions, but also to enable assessment of the operating efficiency of individual transport system components.

In accordance with Journal of Laws of 1999, no. 43, item 430 [21], concerning technical conditions to be met by public access roads and their location in the road network, the following categories of public roads have been functioning in Poland (in parentheses: road classes to which the given road category may be assigned):

- national roads (A, S, GP),
- provincial roads (GP, G),
- district roads (GP, G, Z),
- communal roads (GP, G, Z, L, D).

© Springer Nature Switzerland AG 2019
E. Macioszek and G. Sierpiński (Eds.): Directions of Development of Transport
Networks and Traffic Engineering, LNNS 51, pp. 66–76, 2019.
https://doi.org/10.1007/978-3-319-98615-9_6

Provincial roads perform a very important function, as they supplement the network of national roads. As mentioned in paper [22], there are fewer than 850 provincial roads in Poland with the total length of ca. 28500 km. Every provincial road is designated with three digits on a yellow board. Provincial roads are owned by respective local provincial governments of the given voivodeship (province). This category of roads includes links between towns and roads considered particularly important from the voivodeship's perspective as well as roads which matter for defence purposes, not included in the network of national roads. This article addresses outcomes of a detailed analysis of the General Traffic Measurement results for the years 2000–2015 concerning provincial roads. The analysis has been performed with reference to results pertaining to Opolskie Voivodeship.

## 2   Characteristics of the Provincial Road Network in Opolskie Voivodeship (Poland)

Opolskie Voivodeship is situated in the southern part of the country. Its main centre of administration and the capital city is Opole. In accordance with the data published by the Central Statistical Office (GUS) in 2017 [23], in terms of territory, it is the smallest of all Polish voivodeships with the total area of 9412 km$^2$. This region is also characterised by the lowest population (ca. one million inhabitants). The population density in Opolskie Voivodeship ranges at ca. 106 persons per km$^2$ (ranked 11th in the country). The voivodeship is cut through by the A4 motorway, while the total length of the network of provincial roads is more than 1000 km. The following roads located in Opolskie Voivodeship are categorised as provincial roads [24–27]:

- 382 – Stanowice - Świdnica - Dzierżoniów - Ząbkowice Śląskie - Paczków - state border,
- 385 – state border - Tłumaczów - Wolibórz - Ząbkowice Śląskie - Ziębice - Grodków - Jaczowice,
- 396 – Bierutów - Oława - Strzelin,
- 401 – Żłobizna/road no. 94/- Grodków - Skodroszyce - Pakosławice/road no. 46,
- 403 – Łukowice Brzeskie - road no. 401,
- 405 – Niemodlin - Tułowice - Korfantów,
- 406 – Nysa - Jasienica Dolna - Włostowa,
- 407 – Nysa - Korfantów - Łącznik - road no. 414,
- 408 – Kędzierzyn - Koźle - Gliwice,
- 409 – Dębina - Krapkowice - Strzelce Opolskie,
- 410 – Kędzierzyn-Koźle - Kobylice - Biadaczów - the river Oder - Brzeźce/road no. 408,
- 411 – Nysa - Głuchołazy - state border,
- 413 – Ligota Prószkowska - road no. 429,
- 414 – Wrzoski/road no. 94/Opole - Pruszków - Biała - Prudnik/road no. 40,
- 415 – Zimnice/road no. 45/Rogów Opolski - Krapkowice/road no. 409,
- 416 – Żywocie/road no. 45/Głogówek - Głubczyce - Kietrz - Racibórz,
- 417 – Laskowice/road no. 40/Klisino - Szonów - Szczyty - Racibórz,

- 418 – Reńska Wieś/road no. 45/ - Kędzierzyn-Koźle,
- 419 – Nowa Cerekwia - Niekazanice - Branice - state border,
- 420 – Kietrz - Dzierżysław - Pilszcz - state border,
- 421 – Szczyty - Błażejowice - Nędza,
- 422 – Łany/road no. 421/Dzielnica - Przewóz - the Oder river - Dziergowice,
- 423 – Opole - Krapkowice - Zdzieszowice - Kędzierzyn-Koźle,
- 424 – Gwoździce - the Oder - Odrowąż - Gogolin/road no. 409,
- 425 – Bierawa - Kuźnia Raciborska - Rudy,
- 426 – Zawadzkie - Strzelce Opolskie - Olszowa - Kędzierzyn-Koźle,
- 427 – Droga 45 - Zakrzów - Kochaniec - Roszowice - Dzielnica,
- 428 – Dąbrówka Górna - road no. 45,
- 429 – Wawelno - Komprachcice - Prószków - road no. 45,
- 435 – Opole - Wawelno - road no. 46,
- 451 – Oleśnica - Bierutów - Namysłów,
- 454 – Opole - Pokój - Namysłów,
- 457 – Pisarzowice/road no. 39/Popielów - Dobrzeń Wielki,
- 458 – Obórki - Lewin Brzeski - Skorogoszcz - Popielów,
- 459 – Opole - Narok - Skorogoszcz,
- 460 – Kościerzyce - the Oder - Pawłów - Kopanie - road no. 462,
- 461 – Kup - Jełowa,
- 462 – Stobrawa - the Oder - Kopanie - Łosiów - Krzyżowice,
- 463 – Bierdzany - Ozimek - Zawadzkie,
- 464 – Narok - the Oder - Chróścice,
- 465 – Żelazna - the Oder - Dobrzeń Mały,
- 487 – Byczyna- Gorzów Śląski - Olesno,
- 494 – Bierdzany - Olesno - Wręczyca Wielka - Częstochowa,
- 901 – Olesno - Dobrodzień - Zawadzkie - Wielowieś - Pyskowice - Gliwice/road no. 78.

## 3    Analysis of Results of the General Traffic Measurements Conducted over the Years 2000–2015 in Opolskie Voivodeship

Figure 1 shows changes to the value of annual average daily traffic (AADT) for provincial roads in Opolskie Voivodeship in the years 2000–2015. It is evident that the AADT values grew over the entire period subject to analysis. The minimum value came to 2491 vehicles per day (GTM 2000). The highest traffic volume increase was observed under the General Traffic Measurement (GTM) in 2010 (ca. 18%). The maximum AADT value was reported in the provincial roads of the voivodeship in question in the GTM year 2015 (3309 vehicles per day). The years 2000–2015 saw the traffic volume growing by ca. 33%.

Figure 2 shows a breakdown of the AADT values into individual vehicle groups. With regard to motorcycles (Fig. 2), one should note that the highest increase in the AADT value over the entire period analysed was observed under GTM 2010. This is when the index in question grew by as much as ca. 184%. GTM 2015 also evidenced a result higher than in the previous survey, with the maximum value obtained for the entire period subject to analysis (48 vehicles per day). However, in this case, the increase in the AADT value for motorcycles was only found to range at ca. 13%. This index assumed its minimum value in 2005, when the respective measurement revealed 15 vehicles per day.

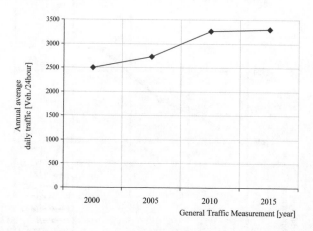

**Fig. 1.** Changes to the annual average daily traffic value for provincial roads in Opolskie Voivodeship in the years 2000–2015

As for the group of passenger cars, one may observe that the AADT value was growing throughout the entire period analysed. The maximum value was observed under the last measurement, and it came to 2778 vehicles per day (increase by 3% compared to the previous survey). The minimum value was reported under GTM 2000 (2029 vehicles per day). From 2000 to 2015, the AADT value for passenger cars increased by 37%.

For light goods vehicles, the changes observed in terms of the AADT values over the entire period of analysis were found to be irregular. Between 2000 and 2010, the increase came to ca. 27%, while a ca. 6% decline in the AADT value for light goods vehicles was observed under the last GTM. The maximum value of this index was reported in 2010, when it came to 240 vehicles per day. The lowest AADT value for light goods vehicles was measured under GTM 2000 (288 vehicles per day).

The predominant trend observed over the whole period of analysis was a decline in the AADT value for heavy goods vehicles without trailers, except for the year 2010, when there was a minimum increase observed (by ca. 4%). The maximum AADT value for heavy goods vehicles without trailers was reported under GTM 2000, and it came to 90 vehicles per day. The minimum value, on the other hand, was 65 vehicles per day, and it was measured in 2015. From 2000 to 2015, the AADT value for heavy goods vehicles without trailers declined in total by ca. 27%.

Another trend observed to dominate over the entire period subject to analysis was also an increase in the AADT value for heavy goods vehicles with trailers, with the total growth of 67%. However, compared to the preceding survey, the increase reported under GTM 2010 equalled 50%. Not until GTM 2015 was a decline in AADT reported for heavy goods vehicles with trailers, its value being ca. 9%. The maximum AADT value for this index was reported under GTM 2010, and it came to 167 vehicles per day. The minimum value, on the other hand, was measured in 2000 at 100 vehicles per day.

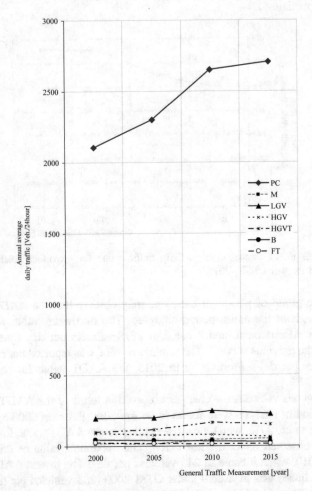

**Fig. 2.** Changes to the annual average daily traffic value for provincial roads in Opolskie Voivodeship in the years 2000–2015 in a breakdown according to vehicle groups (where: PC - passenger cars, LGV - light goods vehicles, M - motorcycles, HGV - heavy goods vehicles, B - buses, FT - farm tractors, HGVT - heavy goods vehicles with trailers)

The AADT values for buses systematically dropped in the years 2000–2015, with the total decline measured over the entire period ranging at ca. 39%. The maximum AADT value for this group of vehicles was reported under GTM 2000, as it came to 44 vehicles per day, while the lowest AADT value for buses was measured in 2015, and it came to 27 vehicles per day.

A drop in the AADT value for farm tractors was also observed over the entire period of analysis. The highest value of this index was reported under GTM 2000, as it came to 23 vehicles per day, while the lowest value was measured in 2015, i.e. 13 vehicles per day. The total drop in the AADT value for farm tractors observed over the years 2000–2015 came to ca. 43%.

Having analysed the road network load induced by AADT (Fig. 3), one can ob served that following the initial drop reported in 2005, the respective value increased in subsequent years. The maximum value of this index was reported under GTM 2015, as it came to 3.56 vehicles per day, while the minimum value was observed under GTM 2005, and it equalled 2.82 vehicles per day. Furthermore, the years 2000–2015 saw a slight total increase in the road network load caused by AADT ranging at ca. 10%.

Figure 4 illustrates the changes to the value of the road network load induced by AADT for provincial roads in Opolskie Voivodeship in the years 2000–2015 in a breakdown into individual vehicle groups. Passenger cars are characterised by the highest value of AADT-induced road network load among all vehicle groups. An increase of ca. 14% was observed in this respect for passenger cars. The minimum value was reported under GTM 2005 (2.36 vehicles per km·day), while the maximum value was measured in 2015, and it came to 2.99 vehicles per km·day.

At the beginning of the period of analysis, on the other hand, there was a slight drop in the AADT-induced road network load observed for motorcycles. The growth phenomenon escalated in 2010, when the index in question came to 0.045 vehicle per km·day (increase by ca. 194% compared to previous measurement). In subsequent years, the AADT-induced road network load was also observed to increase for motorcycles (by ca. 15%). The increase of this index measured over the entire period of analysis came to 146% in total.

Values of the road network load caused by AADT varied for light goods vehicles. A decline of ca. 18% was reported under GTM 2005, as the index came to 0.199 vehicle per km·day, this being the minimum value for the entire period subject to

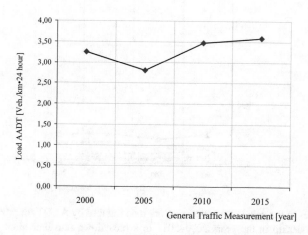

**Fig. 3.** Road network load induced by annual average daily traffic in Opolskie Voivodeship in the years 2000–2015

analysis. The next survey conducted in 2010 revealed the maximum value of road network load, which came to 0.254 vehicle per km·day. GTM 2015 delivered a very similar result in terms of this index compared to the year 2000.

A decline was reported for the road network load caused by AADT in the group of heavy goods vehicles without trailers. Only in 2010 was a minimum increase in the value of AADT-induced road network load reported, as it came to ca. 8% compared to the previous measurement. The highest change to the road network load value was measured in 2005 (ca. 34%). Over the entire period of analysis, the value of this index dropped by 39%.

Another considerably variable index was the AADT-induced road network load attributable to heavy goods vehicles with trailers. It was found to decline by ca. 12% in 2005, when the value came to 0.114 vehicle per km·day, this being the minimum value

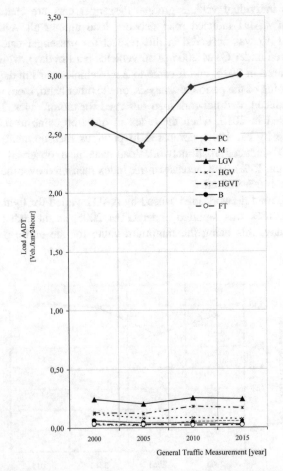

**Fig. 4.** Changes to the value of the road network load induced by AADT for provincial roads in Opolskie Voivodeship in the years 2000–2015 in a breakdown into individual vehicle groups (where: PC - passenger cars, LGV - light goods vehicles, M - motorcycles, HGV - heavy goods vehicles, B - buses, FT - farm tractors, HGVT - heavy goods vehicles with trailers)

for the entire period subject to analysis. The next measurement conducted in 2010 revealed the maximum value of road network load, as it came to 0.176 vehicle per km·day. Over the entire period of analysis, the value of this index increased by 27%.

The values of road network load induced by AADT attributable to buses were gradually declining each time the measurement was taken. The most significant drop was revealed in 2005, as it came to ca. 30%. The highest value was reported in 2000 – it equalled 0.057 vehicle per km·day. The minimum value, on the other hand, was measured in 2015, as it came to 0.029 vehicle per km·day. Throughout the whole period subject to analysis, the AADT-induced road network load due to buses declined by ca. 49%.

The AADT-induced road network load attributable to farm tractors dropped over the entire period of analysis by ca. 53%. The most significant drop was reported in 2005 (ca. 41%). The maximum load value was measured in 2000, as it equalled 0.030 vehicle per km·day. The minimum value of 0.014 vehicle per km·day was obtained in 2015.

Another stage of the analyses covering Opolskie Voivodeship consisted in identifying the provincial roads for which the highest AADT values were established in the 2015 survey. They were provincial roads number 409, 418 and 451. Figure 5 illustrates changes to the AADT values over the years 2000–2015 for the above three roads.

The AADT index for provincial road no. 409 was systematically growing in the years 2005-2010. It came to 5930 vehicles per day in 2015, which ranked this road third over the entire voivodeship in terms of the AADT value. Under the previous GTMs, provincial road no. 409 was ranked fifth in 2000 and sixth in the measurements performed in 2005 and 2010. Between 2000 and 2015, the AADT value for provincial road no. 409 increased in total by ca. 39%.

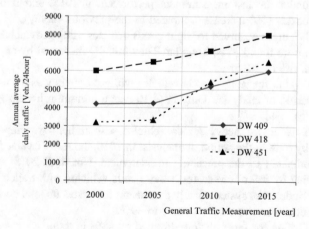

**Fig. 5.** Changes to the AADT value for provincial roads no. 409, 418 and 451 in the years 2000–2015

Similarly systematic was the growth of the AADT index for provincial road no. 418 in the years 2000–2015. The value reported under GTM 2015 was the highest in the voivodeship-wide scale. Previous measurements conducted at provincial road no. 418 also proved that its AADT index was the highest in the region. Its value increased by 32% over the years 2000–2015.

The growth of the AADT index for provincial road no. 451 was also systematic in the years 2000–2015. In 2015, provincial road no. 451 was ranked second in the voivodeship in terms of the AADT value for provincial roads, and it came to 6504 vehicles per day. Between 2000 and 2005, this road was by no means among those with a high AADT index. In this respect, it was ranked eighth and twelfth, respectively. Under GTM 2010, its rating was already the sixth highest in the voivodeship-wide scale. In the years 2000–2015, the AADT value for provincial road no. 451 increased in total by ca. 99%.

## 4  Conclusions

Periodically performed traffic measurements allow to identify changes in volume of traffic on the selected area. It is strong support for local authorities. During subsequent measurements can be seen variability of AADT. Observed variability of AADT results from the growth of motorization as well as the construction of a road network in Poland. Based on results of the analyses addressing the General Traffic Measurement campaign performed in Opolskie Voivodeship in the years 2000–2015, one can formulate the following conclusions:

- the AADT value for provincial roads increased in total by ca. 33%,
- the highest AADT value increase was observed in 2010. This is when this index grew by as much as ca. 184%,
- the AADT value for passenger cars increased by 37%. The maximum value was observed under the last measurement performed in 2015, and it came to 2778 vehicles per day (3% increase compared to the previous survey),
- the AADT changes observed for light goods vehicles using provincial roads subject to analysis were found to be irregular. They initially increased by ca. 27%, only to drop by ca. 6% in 2015,
- the AADT value for heavy goods vehicles without trailers was declining (by 27% over the period of analysis). Only in 2010 was a minimum increase of this index observed (ca. 4%),
- the AADT value for heavy goods vehicles with trailers was reported to have increased by 67% in the period of analysis. In 2010, the AADT value increase came to 50% compared to the preceding measurement. Not until 2015 was an AADT value decline by 9% observed for heavy goods vehicles with trailers,
- the AADT index was systematically decreasing for buses. Its total decrease over the entire period subject to analysis came to 39%,
- the AADT value for farm tractors dropped by 43% in total.

# References

1. Celiński, I.: Using GT planner to improve the functioning of public transport. In: Macioszek, E., Sierpiński, G. (eds.) Recent Advances in Traffic Engineering for Transport Networks and Systems. LNNS, vol. 21, pp. 151–160. Springer, Cham (2018)
2. Celiński, I.: Transport network parametrisation using the GTAlg tool. In: Macioszek, E., Sierpiński, G. (eds.) Contemporary Challenges of Transport Systems and Traffic Engineering. LNNS, vol. 2, pp. 111–123. Springer, Cham (2017)
3. Celiński, I.: Support for green logistics using the GTAlg tool. In: Sierpiński, G. (ed.) Intelligent Transport Systems and Travel Behaviour. AISC, vol. 505, pp. 121–134. Springer, Cham (2017)
4. Sierpiński, G.: Distance and frequency of travels made with selected means of transport - a case study for the upper silesian conurbation (Poland). In: Sierpiński, G. (ed.) Intelligent Transport Systems and Travel Behaviour. AISC, vol. 505, pp. 75–85. Springer, Cham (2017)
5. Sierpiński, G.: Technologically advanced and responsible travel planning assisted by GT planner. In: Macioszek, E., Sierpiński, G. (eds.) Contemporary Challenges of Transport Systems and Traffic Engineering. LNNS, vol. 2, pp. 65–77. Springer, Cham (2017)
6. Pypno, C., Sierpiński, G.: Automated large capacity multi-story garage - concept and modeling of client service processes. Autom. Constr. **81C**, 422–433 (2017)
7. Staniek, M.: Stereo vision method application to road inspection. Baltic J. Road Bridge Eng. **12**(1), 38–47 (2017)
8. Staniek, M.: Road pavement condition as a determinant of travelling comfort. In: Sierpiński, G. (ed.) Intelligent Transport Systems and Travel Behaviour. AISC, vol. 505, pp. 99–107. Springer, Cham (2017)
9. Staniek, M.: Moulding of travelling behaviour patterns entailing the condition of road infrastructure. In: Macioszek, E., Sierpiński, G. (eds.) Contemporary Challenges of Transport Systems and Traffic Engineering. LNNS, vol. 2, pp. 181–191. Springer, Cham (2017)
10. Małecki, K., Wątróbski, J.: Cellular automaton to study the impact of changes in traffic rules in a roundabout: a preliminary approach. Appl. Sci. **7**(7), 1–21 (2017)
11. Małecki, K.: Graph cellular automata with relation-based neighbourhoods of cells for complex systems modelling: a case of traffic simulation. Symmetry **9**(332), 1–23 (2017)
12. Małecki, K.: The use of heterogeneous cellular automata to study the capacity of the roundabout. In: Rutkowski, L., Korytkowski, M., Scherer, R., Tadeusiewicz, R., Lotfi Zadeh, A., Zurada, J.M. (eds.) Artificial Intelligence and Soft Computing. LNAI, vol. 10246, pp. 308–317. Springer, Cham (2017)
13. Macioszek, E., Czerniakowski, M.: Road traffic safety-related changes introduced on T. Kościuszki and Królowej Jadwigi Streets in Dąbrowa Górnicza between 2006 and 2015. Sci. J. Silesian Univ. Technol. Ser. Transp. **96**, 95–104 (2017)
14. Macioszek, E., Lach, D.: Analysis of the results of general traffic measurements in the West Pomeranian Voivodeship from 2005 to 2015. Sci. J. Silesian Univ. Technol. Ser. Transp. **97**, 93–104 (2017)
15. Turoń, K., Czech, P., Juzek, M.: The concept of walkable city as an alternative form of urban mobility. Sci. J. Silesian Univ. Technol. Ser. Transp. **95**, 223–230 (2017)
16. Turoń, K., Golba, D., Czech, P.: The analysis of progress CSR good practices areas in logistic companies based on reports "Responsible Business in Poland. Good Practices" in 2010–2014. Sci. J. Silesian Univ. Technol. Ser. Transp. **89**, 163–171 (2015)
17. Golba, D., Turoń, K., Czech, P.: Diversity as an opportunity and challenge of modern organizations in TSL area. Sci. J. Silesian Univ. Technol. Ser. Transp. **90**, 63–69 (2016)

18. Krawiec, K.: Simulation of technical and economical processes as an initial phase of electric buses fleet implementation to operation in urban public transport company. In: Bąk, M. (ed.) Transport Development Challenges in the Twenty-First Century. SPBE, pp. 193–200. Springer, Cham (2016)

19. Krawiec, K.: Location of electric buses recharging stations using point method procedure. In: Sierpiński, G. (ed.) Intelligent Transport Systems and Travel Behaviour. AISC, vol. 505, pp. 187–194. Springer, Cham (2017)

20. Krawiec, S., Karoń, G., Janecki, R., Sierpiński, G., Krawiec, K., Markusik, S.: Economic conditions to introduce the battery drive to busses in the urban public transport. Transp. Res. Procedia **14**, 2630–2639 (2016)

21. Regulation of the Minister of Transport and Maritime Economy of 2 March 1999 On The Technical Conditions for Public Roads and their Locations. Warsaw: Minister of Transport and Maritime Economy. http://prawo.sejm.gov.pl/isap.nsf/DocDetails.xsp?id=WDU1999 0430430

22. What on the Roads. Provincial Roads in Poland. http://conadrogach.pl

23. Central Statistical Office. Area and Population in Territorial Cross Section in 2017. http:// stat.gov.pl/

24. General Director of National Roads and Motorways. Order No 74 of the General Director of National Roads and Motorways of 2 December 2008 on Numbering Voivodship Roads. https://www.gddkia.gov.pl/userfiles/articles/z/zarzadzenia-generalnego-dyrektor_6335// documents/32_2004.pdf

25. General Director of National Roads and Motorways. Order No 78 of the General Director of National Roads and Motorways of 11 December 2009 Amending the Ordinance on the Numbering of Voivodeship Roads. https://www.gddkia.gov.pl/userfiles/articles/z/ zarzadzenia-generalnego-dyrektor_3184//documents/scan2080.pdf

26. General Director of National Roads and Motorways. Order No. 45 of the General Director of National Roads and Motorways of 17 December 2012 Amending the Ordinance on the Numbering of Voivodeship Roads. https://www.gddkia.gov.pl/userfiles/articles/z/zarzad zenia-generalnego-dyrektor_10385/zarzadzenie%2045.pdf

27. General Director of National Roads and Motorways. Order No 61 of the General Director of National Roads and Motorways of 20 December 2013 Amending the Ordinance on the Numbering of Voivodeship Roads, https://www.gddkia.gov.pl/userfiles/articles/z/zarzadze nia-generalnego-dyrektor_11943/zarzadzenie%2061.pdf

# Quality of Electrical Energy Power Supply of Railway Traffic Control Devices

Jerzy Wojciechowski[1]([⊠]), Łukasz Stelmach[1], Marek Wójtowicz[2], and Jakub Młyńczak[3]

[1] Faculty of Transport and Electrical, Kazimierz Pulaski University of Technology and Humanities in Radom, Radom, Poland
{j.wojciechowski,l.stelmach}@uthrad.pl
[2] Faculty of Computer Science and Mathematics, Kazimierz Pulaski University of Technology and Humanities in Radom, Radom, Poland
m.wojtowicz@uthrad.pl
[3] Faculty of Transport, Silesian University of Technology, Katowice, Poland
jakub.mlynczak@polsl.pl

**Abstract.** The aim of this article was to present an analysis of the quality of electrical energy power supply of railway traffic control devices. The railway traffic control devices belong to a very important group of systems, which role is to assure railway traffic safety. The article focuses mainly on showing problems connected to power supply failure rate and the quality of energy supplied. Within the period of 2 years, a statistical analysis of the power supply failures was performed, also measurements of electrical energy quality parameters in chosen power supply system points were taken. The measurements were performed using the EN50160 norm. In the conclusion, proposals regarding raising reliability and quality of electrical power supply of railway traffic control systems were presented.

**Keywords:** Railway traffic control systems · Electrical power supply
Failure rate · Power supply reliability · EN50160

## 1 Introduction

Railway traffic control is a field of science which main role is to assure railway traffic safety. The safety of people travelling by trains, as well as railway traffic efficiency are assured due to special devices for controlling all rail vehicles [1–12]. The power supply of these devices is assured thanks to systems meeting high technical requirements. High efficiency and reliability of the power supply systems and railway traffic control devices guarantee safe train journeys [3, 4, 13].

Because of the railway traffic control systems and devices' rank in the correct rail traffic operation, their electrical energy power includes basic power supply and backup power. The basic power of the railway traffic control devices is due to the help of special 15 kV power lines, called the lineside power supply. The backup power of the railway traffic control devices constitutes diesel-electric units, which produce electrical energy with parameters similar to the parameters of the power grid.

© Springer Nature Switzerland AG 2019
E. Macioszek and G. Sierpiński (Eds.): Directions of Development of Transport
Networks and Traffic Engineering, LNNS 51, pp. 77–86, 2019.
https://doi.org/10.1007/978-3-319-98615-9_7

Railway traffic control devices are powered with alternating current (AC) voltage, but also with direct current (DC) voltage. Direct current power comes from batteries operating in two systems (uninterruptible and backup) [1].

Railway traffic control devices go with the following electrical ratings:

- AC voltage with the value of 3x400/230 V, converted to other values: 3x230 V, 230 V, 145 V, 130 V, 110 V,
- DC voltage with the value of 24 V, serving for powering relays,
- DC voltage with the value of 24 V, serving for powering signal converters.

The basic way of supplying railway traffic control devices with electricity is the lineside power supply. Within the lineside power supply two power lines operate, one basic and one backup. Switching supply from one line to another and from the lineside power supply to the diesel-electric unit takes place automatically through the Automatic Transfer Switch System. A block diagram of electrical power supply of railway traffic control systems and devices has been presented in the Fig. 1.

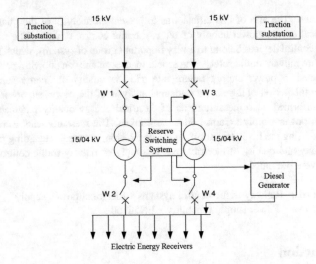

**Fig. 1.** A block diagram of electrical power supply of a railway traffic control system (the lineside power supply and a diesel-electric unit) (Source: [1])

Particular railway traffic control systems, especially computer devices, contain additional, individual backup power supply systems. These are, among others: MUSZ-1E, ELZAS, ZUS, SZS-105, SUZ-1/MONAT/07 supply systems.

One of the most basic factors assuring correct operation of the railway traffic control systems is their power supply. This factor fulfils technical requirements if:

- there is uninterruptible power supply, possible technical problems are removed within the time, in which the railway traffic control device can operate correctly,
- voltage supplied by power supply systems fulfils quality criteria (EN 50160 [14]).

## 2   An Analysis of Failures of Electrical Power Supply of Railway Traffic Control Systems

The presented below analysis of failures of electrical power supply of railway traffic control devices concerns railway lines on a representative fragment of the country, covering about 10% of its area. The analysis was conducted for years 2015–2016. In this period of time damages of systems, elements of railway traffic control systems devices and power supply took place various times. If we assume that failures taking place in 2015 consisted 100%, then in 2016 they consisted 134%.

Failures in electrical power supplies of railway traffic control devices can be divided into 5 types. These are:

- fuse failure in the supply system,
- temporary voltage drop in the power grid of the lineside power supply,
- computer devices failure,
- failures resulting from bad weather conditions,
- damaged elements and components of supply devices.

Fuse failure in the supply system - it was the most frequent failure taking place in the analysed period of time. It occurred 33 times in 2015 and 37 times in 2016. The frequency of this failure is much higher than other failures. There may be many causes of this situation, some of them of a deterministic nature, some of stochastic. These may include: short circuits in AC power cords, poor cords insulation, poor insulation resistance relative to ground or damages caused by maintenance workers.

A momentary voltage drop in the lineside power supply - failures of this kind were significantly rarer than the failures mentioned earlier. In 2015 this breakdown occurred 7 times and in 2016 - 10 times. There are also many causes of this failure, among them are: line-to-line short circuits, earth faults, torn phase wires, insulators' damages, operation faults.

Failures of computer devices (assuring supply to systems responsible for railway traffic control) - these damages appeared sporadically. Such failures pose a direct threat to rail traffic safety and need to be removed within a short period of time. Three breakdowns that took place within two year show that computer devices assuring railway traffic control safety and supplying railway traffic control systems and devices are reliable and guarantee the right level of safety.

Failures resulting from bad weather conditions - the occurrence of this environmental disturbance very often leads to very serious consequences in the operation of the railway traffic control systems. In 2015 only one failure of this type occurred, making this year rather uneventful. Year 2016 was different because of frequent storms. During this period 12 breakdowns took place, they were caused by, among others, gales and lightning discharges.

Damaged elements and components of supply devices - in 2015 there were 10 failures of this kind, in 2016-7. The number of breakdowns may suggest deterioration of technical condition of devices in constant operation. Older devices and railway traffic control systems which are worn out are systematically substituted by new models. The older devices are not in good technical condition, thus the number of failures of this type will maintain on a relatively high level.

The percentage share of the above failures in the analysed area of railway supply in the years 2015 and 2016 is presented in the Table 1 and Fig. 2.

**Table 1.** The number of failures of electrical energy power supply of railway traffic control devices in the years 2015–2016 for the analysed railway area.

| | Type of failure (symbols coherent with the content above) | | | | |
|---|---|---|---|---|---|
| | Fuse failure in the supply system (1) | Temporary voltage drop in the supply network of the lineside power supply (2) | Computer devices failure (3) | Failures resulting from bad weather conditions (4) | Damaged elements and components of supply devices (5) |
| 2015 | 33 | 7 | 1 | 1 | 9 |
| 2016 | 37 | 10 | 2 | 12 | 7 |

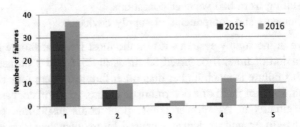

**Fig. 2.** Failures of electrical energy power supply of railway traffic control devices in the years 2015–2016 for the analysed railway area

An interesting issue is the occurrence of failures depending on the season of the year. Such division for the years 2015 and 2016 has been presented in the Table 2.

Spring is the season most abundant in the number of breakdowns (in 2015–23 failures, 2016–24 failures). These breakdowns were connected to, among others, big temperature changes, spring thaws, ground frosts and storms. Summer brings changeable weather that is why breakdowns occurring during this season of the year depend mostly on the weather conditions. In 2015 the occurrence of the weather conditions such as storms and gales was rare, that is why the number of failures was low and amounted to 3. In 2016 there were much more storms and gales, thus the number of breakdowns of railway traffic control systems went up. Failures take place frequently in autumn because during this season heavy rains occur. They contribute to the deterioration of the technical condition of the railway traffic control devices due to high humidity which seriously disrupts the operation of electronic systems. 10 failures that took place in 2015 and 12 failures in 2016 prove that autumn is a dangerous season for the railway traffic control systems. The last, but not least is winter. Low temperatures and heavy snowfalls cause frequent and serious damages. These are mostly torn lineside power supplies and failures of external, non-heated devices.

A crucial factor is the time of fixing a failure of the railway traffic control devices power supply. Figure 3 presents a comparison of amount of failures in 2015, depending on their duration.

From the graph it results that time from the beginning of the breakdown till its fix varied. Most of the technical problems were removed within 0.5–1.5 h, but many of the breakdowns lasted longer, usually up to 3 h. A vast number of failures lasted more than 10 h. It was caused by their type, i.e. a technical problem which did not affect railway safety directly.

**Table 2.** The number of failures of electrical energy power supply of railway traffic control devices (2015–2016) for the analysed railway area, depending on the season of the year.

|        | 2015 | 2016 |
|--------|------|------|
| Spring | 23   | 24   |
| Summer | 3    | 19   |
| Autumn | 10   | 12   |
| Winter | 15   | 13   |

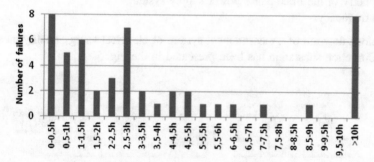

**Fig. 3.** A comparison of amount of failures in 2015 for the analysed railway traffic area, depending on their duration

Figure 4 presents a comparison of amount of failures in 2016, depending on their duration. A vast number of failures occurring in 2016 lasted not more than 0.5 h. Several technical problems were removed within 0.5–2.5 h. There were 7 failures lasting more than 10 h.

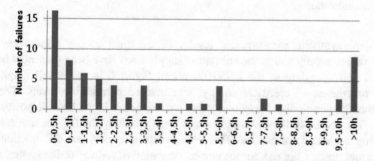

**Fig. 4.** A comparison of amount of failures in 2016 for the analysed railway traffic area, depending on their duration

## 3   An Analysis of Quality Parameters of Electrical Energy in Railway Traffic Control Systems Power Supply

Apart from the failure frequency of railway traffic control devices power supplies, another important factor is the quality of electrical energy supplied [15–20]. Within the conducted analysis of the operation of the power supplies of railway traffic control systems, measurements of electrical energy parameters in the lineside power supply and lines supplying traction substations. The measurements were taken in compliance with the EN 50160 norm, below parameters were taken into account:

- voltage fluctuations,
- voltage drops,
- unpredicted events,
- voltage and current harmonics,
- flicker factor,
- symmetry of the three-phase power supply system,
- grid frequency.

A block diagram of a measurement system of electrical energy quality parameters in a DC traction substation has been presented in the Fig. 5.

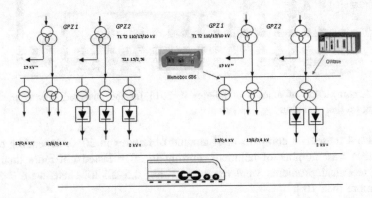

**Fig. 5.** A block diagram of a measurement system of electrical energy quality parameters in a DC traction substation

Chosen, synthetic measurement results of electrical energy parameters in the lineside power supply and in the substation supply lines have been presented below in the form corresponding to the EN50160 norm. Figure 6 is an overall graph of the quality parameters of electrical energy with marked permissible levels. The graph shows that the permissible harmonics value was exceeded in the lineside power supply voltage. The source of these distortions was most probably the DC traction substation generating current deformations. This phenomena is typical for DC traction power supplies and poses a big risk for the correct operation of railway traffic control devices and other systems supplied with the lineside power supply. Table 3 presents measurement data gathered from a chosen lineside power supply. Values exceeding permissible tolerance are marked with light grey and dark grey (EN 50160). With the dark grey colour, exceeded values for 95% $U_N$ measurements are marked, with light grey exceeded values for $U_{N\,max}$.

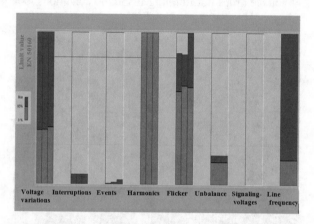

**Fig. 6.** An overall graph of quality parameters of electrical energy supplying the lineside power supply in a chosen DC traction substation (EN50160)

As it has already been mentioned, the cause of incorrect electrical energy quality parameters in the lineside power supply is the operation of rectifier units of the DC traction substation. Because of that, a measurement cycle of electrical energy quality parameters in the power lines supplying traction substations was performed. Synthetic results are presented below.

**Table 3.** Electrical energy quality parameters in a chosen lineside power supply - EN50160.

| Parameter | | Limit Value EN 50160 | Measured 100 % $U_N$ | | Value | Measured 95 % $U_N$ | | Value |
|---|---|---|---|---|---|---|---|---|
| | | | $L_1$ | $L_2$ | $L_3$ | $L_1$ | $L_2$ | $L_3$ |
| **Nominal Voltage $U_N$= 6 kV** | | | | | | | | |
| **Maximum   Value** | % | +10 / +10 | 4.95 | 5.11 | 5.34 | 4.28 | 4.30 | 4.49 |
| $U_N$ **100% / 95%** | kV | 6.6/6.6 | 6.29 | 6.30 | 6.32 | 6.29 | 6.30 | 6.32 |
| **Minimum   Value** | % | -15/-10 | -23.76 | -23.88 | -23.64 | -4.28 | -4.30 | -4.49 |
| $U_N$ **100% / 95%** | kV | 5.1/5.4 | 4.57 | 4.57 | 4.58 | 4.57 | 4.57 | 4.58 |
| **Harmonic Components** | | | | | | | | |
| **13st harmonic $U_N$** | % | 3.0 | 10.82 | 11.02 | - | 6.49 | 6.83 | - |
| **17st harmonic $U_N$** | % | 2.0 | - | - | 8.78 | 0.738 | 0.782 | 5.01 |
| **Flicker** | Plt | 1.0 | 1.039 | 1.029 | 1.258 | 0.738 | 0.782 | 0.767 |

Table 4 presents measurement data gathered from a chosen power line supplying a traction substation. Values exceeding permissible tolerance are marked with light grey and dark grey (EN 50160). With the dark grey colour, exceeded values for 95% $U_N$ measurements are marked, with light grey exceeded values for $U_{Nmax}$. In this case, many events and voltage drops not fulfilling EN50160 conditions took place (Fig. 7).

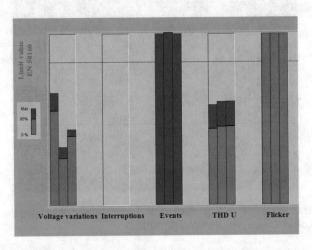

**Fig. 7.** An overall graph of quality parameters of electrical energy supplying a chosen DC traction substation (EN50160)

**Table 4.** Electrical energy quality parameters in a chosen DC traction substation - EN50160.

| Parameter | | Limit Value EN 50160 | Measured Value 100% $U_N$ | | | Measured Value 95% $U_N$ | | |
|---|---|---|---|---|---|---|---|---|
| | | | $L_1$ | $L_2$ | $L_3$ | $L_1$ | $L_2$ | $L_3$ |
| Nominal Voltage $U_N = 30$ kV | | | | | | | | |
| Maximum Value $U_N$ 100% / 95% | % | +10 / +10 | 7.88 | 4.09 | 5.33 | 6.63 | 3.32 | 4.85 |
| Minimum Value $U_N$ 100% / 95% | % | -15/-10 | 2.40 | -0.87 | 0.50 | 2.40 | -0.87 | 0.50 |
| Dips | - | 100 | 704 | 121 | 225 | - | - | - |
| Harmonic Components | | | | | | | | |
| THD | % | 8.0 | 5.63 | 5.79 | 5.81 | 4.28 | 4.4 | 4.42 |
| Flicker | Plt | 1.0 | 3.785 | 2.762 | 3.074 | 1.741 | 1.408 | 1.383 |

What results from the conducted measurements is that, in many cases, electrical energy quality parameters do not comply with values in the EN50160 norm. The reason of this situation is mainly the specifics of the traction substation operation: their non-lineal character and a substantial variability of traction load. Such work conditions cause a deterioration of power conditions of the lineside power supply, which is the basic source of power for railway traffic control systems.

# 4 Conclusions

The aim of this article was to present an analysis of power supplies of railway traffic control devices. These systems belong to a very important group of devices assuring railway traffic safety. The emphasis was mainly put on showing reliability issues of power supply systems and energy parameters supplied by them. Because of that, a two-year statistics of power supply systems' failures and measurements of electrical energy quality parameters in the chosen points of railway traffic control systems power supply were conducted.

From the conducted research it results that the biggest group of breakdowns of power supplies of railway traffic control systems are group 1 and 5, which are fuse failure and damaged elements and components of supply devices. Where it comes to the fuse failure, usually it is caused by the device's internal damage, damage of elements and components. This fact shows how important it is to substitute old, worn-out devices with new ones, equipped with developed internal diagnostics. Such action will substantially limit failures of this kind. The second most frequently occurring technical problems were failures related to bad weather conditions. The environmental factor is hard to eliminate but, in this case, one should take up efforts to shorten the time of repairing the damage. A special type of threat are failures connected to storms and lightning discharges. In this case one should sensitize railway services to the necessity of knowing weather forecasts. A perfect tool for that are widely available computer program, e.g. EUCLID.

From the conducted measurements of electrical energy quality parameters in the lineside power supply and power lines supplying traction substations it results that in case of the DC traction, energy quality is an uncertain factor which needs constant monitoring. Non-lineal character of rectifier units and a high variability of traction load are elements deciding about the deterioration of energy quality supplied to railway traffic control systems. Another issue is the sense of using the EN 50160 norm with reference to powering systems especially sensitive to the quality of energy, which are railway traffic control systems. Separate regulations concerning the power supply of systems particularly sensitive should be implemented.

Apart from the occurring problems, contemporarily used power supplies of railway traffic control systems are able to assure uninterruptable, reliable and efficient power supply, which is connected to the railway safety.

# References

1. Mikulski, A., Tajer, T.: Maszyny i Urządzenia Elektryczne w Automatyce SRK. Wydawnictwo Komunikacji i Łączności, Warszawa (1989)
2. Łukasik, Z., Ciszewski, T., Młyńczak, J., Nowakowski, W., Wojciechowski, J.: Assessment of the safety of microprocessor-based semi-automatic block signalling system. In: Macioszek, E., Sierpiński, G. (eds.) Contemporary Challenges of Transport Systems and Traffic Engineering. LNNS, vol. 2, pp. 137–144. Springer, Switzerland (2017)
3. Łukasik, Z., Nowakowski, W.: Application of TTCN-3 for testing of railway interlocking systems. In: Mikulski, J. (ed.) Transport Systems Telematics. CCIS, vol. 104, pp. 447–454. Springer, Berlin (2010)

4. Ciszewski, T., Nowakowski, W.: Interoperability of IT systems in the international railways. In: 16th International Scientific Conference on Globalization and its Socio-Economic Consequences. Proceedings Part I, pp. 312–320. University of Zilina, Zilina (2016)
5. Białoń, A., Gradowski, P., Iwański, R.: ERTMS Regional - The Control Command System Proposal for Low Density Lines. Advances in Transport Systems Telematics. WKiŁ, Warsaw (2008)
6. Lewiński, A., Perzyński, T., Bester, L.: Computer aided safety analysis of railway control systems. J. KONBiN 2(26), 137–150 (2013)
7. Lewiński, A., Perzyński, T.: The reliability and safety of railway control systems based on new information technologies. In: Mikulski, J. (ed.) Transport Systems Telematics. CCIS, vol. 104, pp. 427–433. Springer, Berlin (2010)
8. Lewinski, A., Trzaska-Rycaj, K.: The safety related software for railway control with respect to automatic level crossing signaling system. In: Mikulski, J. (ed.) Transport Systems Telematics. CCIS, vol. 104, pp. 202–209. Springer, Berlin (2010)
9. Liu, B., Ghazel, M., Toguyeni, A.: Model-based diagnosis of multi-track level crossing plants. IEEE Trans. Intell. Transp. Syst. 17(2), 546–556 (2016)
10. Liu, M., Yan, X., Sun, X., Dong, W., Ji, Y.: Fault diagnosis method for railway turnout control circuit based on information fusion. In: Information Technology, Networking, Electronic and Automation Control Conference Proceedings, Chongqing, pp. 315–320. IEEE (2016)
11. Liu, P., Yang, L., Gao, Z., Li, S., Gao, Y.: Fault tree analysis combined with quantitative analysis for high-speed railway accidents. Saf. Sci. 79, 344–357 (2015)
12. Martorell, S., Soares, C.G., Barnett, J.: Safety, Reliability and Risk Analysis: Theory, Methods and Applications. CRC Press, Taylor & Francis Group, Boca Raton, London, New York, Leiden (2009)
13. Łukasik, L., Ciszewski, T., Wojciechowski, J.: Power supply safety of railway traffic control systems as a part of international transport safety. In: 16th International Scientific Conference on Globalization and its Socio-Economic Consequences. Proceedings Part IV, pp. 1212–1219. University of Zilina, Zilina (2016)
14. Klajn, A., Bątkiewicz-Pantuła, M.: EN 50160. Voltage Characteristics of Electricity Supplied by Public Electricity Networks. European Copper Institute, Woluwe, Saint Pierre (2013)
15. Saied, M., Al-Shaher, M.: Harmonic distortion assessment and minimization for railway systems. Electr. Power Compon. Syst. 37, 832–846 (2009)
16. Vujatovica, D., Koob, K.L., Eminb, Z.: Methodology of calculating harmonic distortion from multiple traction loads. Electr. Power Syst. Res. 138, 165–171 (2016)
17. Gunavardhini, N., Chandrasekaran, M., Sharmeela, C., Manohar, K.: A case study on power quality issues in the indian railway traction sub-station. In: 7th International Conference on Intelligent Systems and Control, Coimbatore, pp. 7–12. IEEE Press (2013)
18. Xiangjing, Z., Baohua, W.: The power quality analysis of traction power supply system. In: IEEE Chinese Control and Decision Conference, Nanjing, pp. 6313–6318. IEEE Press (2014)
19. Lei, X., Yang, G., Liu, W., Liu, C., Wang, Y., Huang, K., Deng, Y.: Feasibility study on the power supply of DC 3-kV system in urban railway. In: Jia, L., Liu, Z., Qin, Y., Ding, R., Diao, L. (eds.) Proceedings of the 2015 International Conference on Electrical and Information Technologies for Rail Transportation. LNEE, vol. 377, pp. 309–318. Springer, Berlin (2016)
20. Sadeghi, J., Masnabadi, A., Mazraeh, A.: Correlations among railway turnout geometry, safety and speeds. In: Proceedings of the Institution of Civil Engineers-Transport, vol. 169, no. 4, pp. 219–229 (2016)

# Development of the Odra River in the Context of Its Use for Conveying of Bulk Materials with Intermodal Transport Technologies

Sylwester Markusik and Aleksander Sobota[✉]

Faculty of Transport, Silesian University of Technology, Katowice, Poland
{sylwester.markusik,aleksander.sobota}@polsl.pl

**Abstract.** Demand for transport of bulk materials in the north-south relation and transport forecasts on this route, justify the launch of intermodal connection to customers located in the south of Europe, using the Baltic Adriatic Transport Corridor (CETC), the AGN network and the Oder waterway (E30). Transport of valuable bulk cargoes (coking coal, coke, fertilizers), exported from the Upper Silesia area, south to river, rail and sea transport to customers located in SouthEastern Europe, and even further, in Egypt, India, and in China (coking plants, steel mills).

**Keywords:** Inland waterways · Inland navigation · Intermodalism
Transport corridors · Innovative transport technologies · Transport costs

## 1 Introduction

The growing demand for freight transport in the north-south relation and forecasts of transport in this direction, in particular bulk materials, justify the launch of a water intermodal connection using the Central European Transport Corridor (CETC) and the AGN network using the Oder Waterway. On the intermodal (container) transport market, the most important will be valuable bulk cargo (coking coal, coke, granular metallurgical and chemical materials), transported from areas adjacent to the Oder, most often to recipients located in the south of Europe (assuming the use of the future Canal Odra - Danube - Elbe), but also in North Egypt, India and China (coking plants, steelworks). Not without significance is the fact that the Czech Republic in 2023 will definitely abolish coal mining and all supply of coal and coke will come from Polish mines, and the best way to supply Czech steelworks (e.g. Steelworks in Trzyniec) will be the use of the Upper Oder.

Planned exploration of the Odra along its entire length up to the 4th navigability class, suggests an analysis of its use for ecological and economically advantageous intermodal freight, in place of the conventional technology used today, where the material is transported in bulk (e.g. from mines or coking plants from Silesia to ports: Gdansk or Szczecin - Świnoujście, coal wagons and then bulk carriers around Europe to the south or east ports).

Intermodal transport of bulk materials in containers, especially large quantities (several million tons a year), can be competitive in relation to the conventional

E. Macioszek and G. Sierpiński (Eds.): Directions of Development of Transport
Networks and Traffic Engineering, LNNS 51, pp. 87–97, 2019.
https://doi.org/10.1007/978-3-319-98615-9_8

(currently used). However, in order to be competitive with conventional transport, it requires, using traditional reloading technologies, large masses to be transported, with relatively large transport distances, and hence an appropriate scale. Experience from the implementation of intermodal technologies shows that economies of scale in transport do not always result in lower total costs [1]. In Poland, intermodal transport is not yet competitively priced compared to conventional transport. Therefore, the question arises - how to reduce these costs?

## 2    Transport Chain for the Export of Valuable Bulk Materials - A Concept

The concept of a transport chain for the export of valuable bulk materials from Poland in the south consists in using the Transport Corridor Baltic - Adriatic or Central European Transport Corridor (CETC), via the E30 waterway (called the Odra Waterway in Poland). On its southern part, it covers the Upper Oder, starting from Opole to the planned canal connecting the Odra with the Elbe and the Danube (ODL) (Fig. 1).

**Fig. 1.** The connection of the Danube - Oder - Elbe rivers with the ODL canal (Source: own on the basis of [7])

These plans are in line with the AGN convention (adopted by Poland only in 2017), in which the main waterways of international importance have been designated on the map of Europe. In the transport chain, the construction of additional handling terminals (hubs) should be considered, which will increase transport accessibility for individual links in this chain.

The development of inland waterways in Poland, in the perspective of 2030, assumes the construction of the Koźle - Ostrava section on the Oder Waterway, as part of the Oder Danube-Elbe (ODL) connection. This idea may be of great importance for Poland as it will enable the introduction of innovative transport technology and popularize intermodal transport of bulk materials in folding containers, increasing the use of river waterways and railways, relieving (especially in terms of $CO_2$ emissions) crowded sea routes around Europe [2–6].

Not without significance is also the improvement of quality (even up to 50%) of coal and coke transported in containers, compared to its transport by coal wagons and the necessity of several trans-shipments at terminals. Then grains crumble and reduce its granulation, and thus its thermal value. Already today, some recipients of Polish coke require its delivery in closed containers, e.g. Steelworks Loeben (Austria) or Metallurgical Plant in Dunkirk (France).

The main purpose of the intermodal transport technology for bulk materials is to develop innovative management methods and to solve advanced planning and modeling of infrastructure performance in intermodal transport. These plans were presented in the Resolution of the Council of Ministers of the Republic of Poland; Journal of Laws No. 79 of June 14, 2016.

The purpose of the new technology of transporting bulk materials on the Oder Waterway is also:

- development of accompanying infrastructure necessary to introduce intermodal transport technologies, through the creation of transshipment terminals (logistic centers),
- improvement of railway connections in the TEN-T network, with high capacity along the north-south transport axis,
- raising the navigability of the waterway to a minimum of class IV (365 navigable days in a year),
- development of the environmental planning system, as a support for "clean" infrastructure projects ("pro-ecological transport" - the so-called "Green Corridor").

## 3  Generators of Demand for Transport by the Oder Waterway

Inland waterways play a major transport role in the European Union, linking land and sea economies. The Oder is a communication artery (part of the TEN - T network), and combines numerous industrial districts and agricultural lands, also located in the territories of the neighboring states of Poland. The areas in the south and north of the river are characterized by a rich industrial infrastructure, and in the middle course - by agriculture. These are highly urbanized areas, and economic indicators there are higher than national averages. Hence the possibility of using the Oder as a transport axis in the north-south direction to intensify freight transport in both directions, which will contribute to deepening the economic cooperation of countries in the planned area of the "Tri-Sea".

In Poland, the production of a significant part of valuable bulk materials (coking coal, coke) is located in the area directly adjacent to the Upper Oder (from Opole to the border with the Czech Republic), while the demand of the domestic metallurgical market for these products does not exceed 50%. In 2014, 9.9 million tons of coking coal were extracted in mines in Poland, which in 50% is processed into coke in own coking plants.

Poland until 2014 was the world's largest coke exporter (about 6 million tons), but has now lost its position as a world leader in coke exports (for China), due to the decline in its prices on global markets caused by the abolition of export duties and licenses (by GATT) and reduction of freight rates. On the other hand, in the coking coal structure, Poland exports only 2.2 million tons in 2015, which is a small share (when coal is extracted this year 13 million tons), compared to exports of these coals by Australia (30.5%), USA (18%) or Canada (10%), with its total consumption in the world equal to 325 million tons [9].

The position of Polish coking coal on global markets reduces the high costs of its extraction and transport (including freight). However, the advantage of Poland is that it has the youngest coke batteries (about 15 years in relation to about 24 years in the world), which allows reducing the cost of coke production and offering it on world stock exchanges (ARA Index- Euro Raiting and RB Index- Global Coal Index) at prices lower than the competition (Fig. 2).

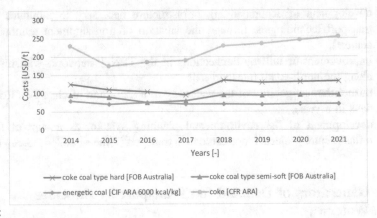

where:
FOB - Free on board,
CIF - Cost, insurance and freight,
CFR - Cost and freight.

**Fig. 2.** Price trends for coke and coals on global markets (Source: own study based on [8] of 2015)

As a prerequisite for the development of Polish coking coal and coke exports, it is possible to reduce its price and adjust it to the current level on the world stock exchanges (ARA and RB) and improve its quality (delivery in containers). It can be assumed that the competitiveness of Polish coal and coke on global exchanges will depend largely on the amount of transport costs (Fig. 2).

In order to achieve a reduction in internal transport costs, it is necessary to shorten the currently used routes of coke/coal export from Silesia to recipients and to use cheaper ways of its transportation (inland water transport). Particular emphasis should be placed on making our carbon products more attractive to existing consumers

(Table 1), due to the currently high activity of Chinese exporters on European and world markets.

**Table 1.** Main export directions of coke from Poland in 2013–2014. (Source: own on the basis of [9], 2015]).

| Direction of export | Number of tones per year [mln. tons] | | Percentage share [%] | |
|---|---|---|---|---|
| | 2013 | 2014 | 2013 | 2014 |
| Germany | 1.61 | 1.63 | 29.27 | 27.16 |
| Italy | 0.484 | 1.17 | 8.80 | 19.5 |
| Austria | 0.963 | 0.832 | 17.5 | 13.86 |
| Romania | 0.689 | 0.645 | 12.5 | 10.75 |
| Czech Republic | 0.345 | 0.259 | 6.25 | 4.32 |
| Slovakia | 0.098 | 0.151 | 1.78 | 2.5 |
| Egypt | 0.251 | 0.26 | 4.56 | 4.33 |
| India | 0.211 | 0.128 | 3.85 | 2.13 |
| Others | 0.85 | 0.925 | 15.49 | 15.45 |
| Export in the south direction | 3.04 | 3.445 | 54.5 | 57 |
| Total | 5.5 | 6.0 | 100 | 100 |

Table 1 shows that more than half of Polish coke exports in 2013–2014 were located south of Polish borders, hence great opportunities to shorten the transport route to customers using the Upper Oder and the ODL Canal in the Czech Republic.

# 4 Intermodal Transport of Bulk Materials with the Use of Innovative Transport Technology

## 4.1 General Rules for the Intermodal Transport of Bulk Materials

Currently, transport of bulk materials (both in road and rail transport) takes place only in open wagons or containers, which promotes dusting during transport and worsens the quality of transported material, associated with the necessary reloading and storage at any change of means of transport. A large number of coal/coke trans-shipments, increases transport time, lowers the efficiency of the transport network, reduces granulation and thermal value and has an adverse environmental impact (dusting).

In intermodal transport, combined transport is often used, where the integrated unit load (container), on the main part of the route, is transported (without reloading its contents) between main terminals (hubs) by rail, inland or sea navigation. Its delivery and transfer (to/from the main terminal) takes place by another means of transport (Fig. 3). The main terminals (hubs) have high efficiency, but their number in the system is smaller [7, 10].

For the intermodal transport of bulk materials, especially valuable ones, containers of a special construction can be used, while maintaining their universal (especially dimensional) qualities. Because in most cases the containers return, they will be unloaded (empty), the construction should enable their assembly, to the form in which the four complexes have the volume and form of one universal container 1C (1 TEU) [5, 6].

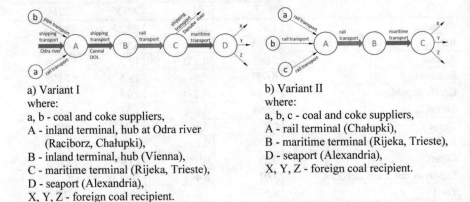

a) Variant I
where:
a, b - coal and coke suppliers,
A - inland terminal, hub at Odra river
   (Raciborz, Chałupki),
B - inland terminal, hub (Vienna),
C - maritime terminal (Rijeka, Trieste),
D - seaport (Alexandria),
X, Y, Z - foreign coal recipient.

b) Variant II
where:
a, b, c - coal and coke suppliers,
A - rail terminal (Chałupki),
B - maritime terminal (Rijeka, Trieste),
D - seaport (Alexandria),
X, Y, Z - foreign coal recipient.

**Fig. 3.** Variants of delivery of bulk materials from the Upper Silesia area to the recipient in Alexandria – (a) Option I, (b) Option II

The new type of folding container for the transport of valuable bulk materials should have the following characteristics:

- size 1C (1 TEU), dimensions and features fully corresponding to ISO containers,
- purpose: transport of bulk materials,
- loading: filled from the top,
- unloading: hinged doors on both front walls,
- container for the return transport time (if empty - without load), should allow its submission.

## 4.2  Intermodal Transport of Bulk Materials Using Innovative Transport Technology - Characteristics of Variants

The following is an example of an analysis of economic and ecological efficiency when transporting bulk materials from Upper Silesia to Egypt (to the Port of Alexandria). The recipient is located in Middle Egypt (Steelworks and Al - Nasr Coking Plant, approx. 250 km from the port of Alexandria), for three different transport versions (versions I–III).

Version I - is an innovative technology of intermodal combined transport, valuable bulk materials, which uses inland water transport (Oder River, ODL Canal and Danube River), rail transport (to the North Adriatic ports) and sea (to D-port in Alexandria) (Fig. 3).

In this technology, the stock is assembled on the shoulders in the main container terminal (A) in southern Poland (near Raciborz or Chałupki) and runs directly between the sender and the indirect recipient via waterway (terminal B - river port Vienna), omitting the operation of dispatching, reloading or additional picking. For recipients located in the area of the river Danube or the Black Sea (e.g. Serbia, Romania, Ukraine, Bulgaria), inland waterway transport may be continued. However, for further recipients from the southern directions (Egypt, India, China), containers in terminal B are transshipped to railway platforms, operating as shuttle trains, to C terminals over the Northern Adriatic (Trieste, Rijeka, Koper), from where they are transported to recipients (terminal D) by sea.

In version II it is assumed to use the technology of a shuttle train, where the train is assembled in the main container terminal (A) and runs directly between the sender and the indirect recipient (sea terminal B), with the intervention, reloading or additional picking operations. From where the sea transport is transported to the recipients (terminal C) by sea (Fig. 4).

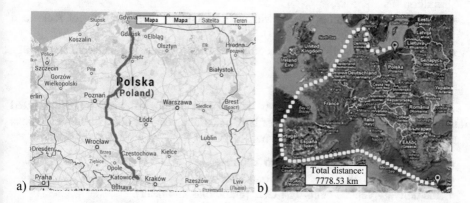

**Fig. 4.** Version III: a- rail route Silesia - Gdansk; b - sea route from Gdansk to Alexandria (Source: own on the basis of [12])

It is also an intermodal technology, applicable without waiting for the revitalization of shipping on the Upper Oder and for the construction of an ODL canal. Sea transport starting from Terminal B takes place further to the recipients by sea according to version I. The system of shuttle trains in intermodal transport of loose materials in TEN-T transport corridors consists in:

- compilation of compact train sets with containers, transporting bulk cargo on a fixed route, e.g. from Silesia to the sea port on the North Adriatic (Trieste, Koper, Rijeka),
- unchanged composition and number of wagons on the train,
- a minimum of 5 train journeys per week,
- the lack of possibility of dispatching the train along the route and at the points of sending and receiving,

- acquisition by the operator of transport from the owner of railway infrastructure (in individual countries in the transport corridor) of traction services, taking full risk of loading the train.

Version III of bulk materials export from Silesia to the recipient located south of Polish borders (conventional, currently used) assumes that the material is transported in bulk (from mines or coking plants) from Silesia to ports: Gdansk or Szczecin - Świnoujście, coal wagons and then by bulk ships around Europe to the southern ports [11] (Fig. 4).

### 4.3    Intermodal Transport of Bulk Materials with the Use of Innovative Transport Technology - Variants Analysis

The analysis compared the following parameters for all three versions:

- length of the route,
- calculated transport costs (internal), with an annual contract for the transport of 1 million tons of coke,
- average transport times (net - without transloading time, stopping at locks and storage at terminals),
- delivery costs of Polish coke (according to CFR - delivery to the recipient's port - Alexandria),
- costs of ecological burden.

In determining the external costs of transport (ecological load) in the analyzed transport versions of 1 million tons of bulk material a year, the costs of $CO_2$ emissions were taken into account, assuming that [6]:

- barge emits 33.4 $CO_2$ [gram/tono kilometer],
- the train emits 48.1 $CO_2$ [gram/tono kilometer],
- the delivery vehicle emits 164 $CO_2$ [gram/tono kilometer].

The $CO_2$ emission in inland and rail water transport is several times smaller than in road transport. External transport costs require reference to the cost of the primary energy source. The original source of energy is one that comes directly from nature. It is, for example, coal, oil, gas, wind or solar radiation (but not electricity). The external costs of transport also include the accident rate and the intensity of the noise emitted, but comparing these factors for inland waterway, rail and sea transport does not make sense. Table 2 presents the volume of $CO_2$ emissions when transporting 1 million tons of loose materials annually. The volume of this emission was calculated on the basis of input data (length of the route, number of transports per year, cargo carried by one transport) with assumptions concerning unitary $CO_2$ emission. The calculations were made using the formula:

$$E_t = E_u \cdot N_{ty} \cdot C \cdot L \qquad (1)$$

where:

$E_t$ - total emission [tons],
$E_u$ - unit emission [grams/tons·km],
$N_{ty}$ - number of transport per year [for the mass of 1 mln tons/year],
$C$ - capacity [tons],
$L$ - length of the route [km].

**Table 2.** Total emission of $CO_2$ in rail, maritime and inland shipping transport. (Source: own on the basis of [13]).

| Total emission of $CO_2$ by branches of transport | Version I intermodal - inland shipping | Version II intermodal - rail | Version III conventional |
|---|---|---|---|
| | Katowice - Vienna - Trieste - Alexandria | Katowice - Trieste - Alexandria | Katowice - Gdansk - Alexandria |
| Rail transport [t] | 10482 | 20485 | 12100 |
| Maritime transport [t] | 38930 | 38930 | 132226 |
| Shipping transport [t] | 13950 | - | - |
| Total [t] | 63362 | 59415 | 144326 |

**Table 3.** Measures for variants comparison.

| Measures for variants comparison | Version I intermodal - inland shipping | Version II intermodal - rail | Version III conventional |
|---|---|---|---|
| Total length of route [km] | 3320 | 3230 | 8350 |
| Total time of transport (net) [h] | 84 | 120 | 217 |
| Ecological costs [USD/t] | 0.75 | 1.1 | 1.25 |
| Production costs of 1 ton of coke in Poland [USD][a] | 97.5 | | |
| Delivery costs of 1 ton of Polish coke to suppliers - CFR [USD][c] | 187.5 | 197.5 | 237.5 |
| Price of 1 ton of coke at stock exchange in Rotterdam - ARA [USD][b] | 220 | | |
| Export profitability of Polish coke [USD/t] | +33.5 | +23.5 | −17.5 |

where:

(a) - [14] (06.05.2017),

(b) - as of 06.05.2017,

(c) - charges for: terminal storage, customs clearance, insurance and administration were not included.

In contrast, Table 3 presents measures for comparing defined variations of delivery of Polish coke/coal to the recipient.

Table 3 shows that the introduction of intermodal transport using cheaper branches (inland water transport) and shorter transport distances (the transport route for version I is 2.5 times shorter than version III) leads not only to shortening the transport time (also 2.5 fold), but above all contributes to obtaining a positive balance of trade in coke and metallurgical coal with contractors located to the south of the Polish borders.

One should also consider the possibility of using version II as a temporary one, until the inland waterway from Poland is cleared southwards, which will most probably only be around 2030. Version II also gives a positive balance in trading in valuable bulk materials with countries south of the borders Polish. However, it requires investing funds in the launch of a new intermodal technology, based on specialized containers for the transport of bulk materials.

# 5  Conclusions

- Inland waterway transport on the Odra River should become an important element of the transeuropean transport corridor (north-south), enabling close connection with the Polish maritime economy. The basic requirement for the Baltic - Danube water connection is the construction of the Oder - Danube - Elbe Canal and the fulfillment of the at least IV navigability class for the Odra Waterway. This will trigger increased and sustainable freight transport in both directions of the Oder.
- The intermodal transport of bulk materials is a significant alternative to conventional transport, offering lower prices for transport services and a higher level of services than in other transport technologies.
- Even if we assume that the data accepted in the calculations (especially the internal costs of transport) are subject to errors, resulting from inaccurate (or unreliable) information collected in logistics companies, the order of the results obtained is clearly important when comparing route variants, which are clearly more beneficial for the intermodal route.
- It should be realized that intermodal transport of bulk materials from Poland, in the north-south direction, will be basically devoid of the possibility of transporting cargoes on the return route, and the containers will then be transported empty (until the Oder-Danube-Łaba Canal starts).

# References

1. Tschirner, P.: Intermodality. A Contribution Towards a Sustainable Development and Environment. World Road Association, Helsinki (2000)
2. International Union of Railways: Combined Transport in Europe. Report 2017. International Union of Railways, Paris (2017)
3. European Court of Auditors: Inland Waterway Transport in Europe: No Significant Improvements in Modal Share and Navigability Conditions Since 2001. Special report, European Court of Auditors, Luxembourg (2015)

4. Kałuża, T., Radecki-Pawlik, A., Szoszkiewicz, K., Plesiński, K., Radecki-Pawlik, B., Laks, I.: Plant basket hydraulic structures as a new river restoration measure. Sci. Total Environ. **627**, 245–255 (2018)
5. Markusik, S., Gąska, D., Mateusiak, P.: Analysis of the possibilities of using VI corridor of transport for the alternative export of bulk cargo from Upper Silesia (Poland) to the Mediterranean Countries. Logist. Transp. **27**, 33–41 (2015)
6. New UE Freight Corridors in the Area of the Central Europe. https://www.port.venice.it/files/page/studiosonoraco2.pdf
7. Markusik, S., Fellner, A., Mikulski, J.: Infrastruktura Logistyczna w Transporcie. Tom III. Wydawnictwo Politechniki Śląskiej, Gliwice (2013)
8. Metallurgical Coce Market Outlook. https://www.smithersapex.com/market-reports/metallurgical-coke-market-price-2017
9. Global Trade Atlas. https://www.gtis.com/gta
10. Park, Y.A., Medda, F.: Hub status and indexation of container ports. Asian J. Shipp. Logist. **31**(2), 253–272 (2015)
11. Pluciński, M.: Możliwość Wykorzystania Transportu Wodnego Śródlądowego w Obsłudze Zespołu Portowego Szczecin-Świnoujście. PTE, Szczecin (2016)
12. Google Maps. http://www.googlemaps.pl
13. European Environment Agency. http://www.eea.europa.eu
14. Newspaper. http://info.wyborcza.biz/szukaj/gospodarka/cena+koksu

# Traffic Engineering as Support for Development of Transport Networks and Systems

# Characteristic Parameters Model of Traffic Flow in Ring Expressway Based on Physical Attributes

Hua-lan Wang$^{(\boxtimes)}$, Jia-ying Xu, and Shu-mei Ma

School of Transportation, Lanzhou Jiaotong University, Lanzhou, China
wanghualan126@126.com, 956005349@qq.com,
752066154@qq.com

**Abstract.** In order to ease the traffic pressure of the city center and to solve the complex traffic function demand. Based on the quantitative analysis of the traffic flow characteristics of the Expressway, the paper puts forward the idea of deducing the characteristic parameters of the traffic flow directly with the physical characteristic factors, and establishes the characteristic parameter model of the traffic flow based on the physical attributes. Taking the Xi'an Ring Expressway as an example, the calibration of characteristic parameters of traffic flow based on physical attribute is carried out and compared with the traditional traffic flow characteristic parameter calibration method. The results show that the characteristic model of the traffic flow of the Ring Expressway based on the physical attributes reduces the dependence on the measured data and it can better describe the expressway traffic flow characteristics.

**Keywords:** Traffic engineering · Ring Expressway
Traffic flow characteristic parameters · Physical characteristics factor
Physical attributes

## 1 Introduction

In Fangxuan's paper [1], with the development of metropolitan traffic network construction, the Ring Expressway has become an important part of the metropolitan traffic network as a circular traffic around the city. The dynamic traffic situation assessment of ring expressway is one of the important basis for the overall traffic network planning and traffic management of the metropolitan area. The dynamic traffic flow model is the basis of dynamic traffic situation assessment, and the reasonable calibration of dynamic traffic flow characteristic parameters is the key to accurate dynamic traffic assessment.

In Jiang and Wu paper [2], traffic flow model means that describes the basic nature of traffic and reveals the basic motion rules of traffic flow, which mainly describes the relationship between speed, flow rate and density. The relationship between speed, flow and density of traffic flow model is analyzed, and the key control points of the curve are studied. In many traffic flow models, the Greenshields [3] model is the most classical macro traffic flow model that describes the relationship between flow, speed, and density, but the Greenshields model assumes that the speed-flow curve is a parabola and the underlying data is not from the expressway. The Pipes model [4] is the most

© Springer Nature Switzerland AG 2019
E. Macioszek and G. Sierpiński (Eds.): Directions of Development of Transport
Networks and Traffic Engineering, LNNS 51, pp. 101–113, 2019.
https://doi.org/10.1007/978-3-319-98615-9_9

classical micro-traffic flow model, but the Pipes model assumes that the speed is not related to the flow in the non-congested condition. Therefore, the Greenshields model and the Pipes model [5] have defective for the description of the traffic flow characteristics of the Ring Expressway. In 1995, Van Aerde proposed a classical four-parameter single-structure model that encompasses the macroscopic Greenshields model and the microscopic L.A. Pipes model, and avoids the drawbacks of the two models. In Wudi's paper [6], compared with the previous two models, the Van Aerde model [7] is more suitable for describing the traffic flow characteristics of the expressway.

To obtain the Van Aerde model of the Ring Expressway, four traffic flow characteristic parameters such as free flow speed (FS), critical speed (SC), jam-up density (JD) and traffic capacity (CAP) are calibrated [8]. The conventional method of calibration of traffic flow characteristic parameters is based on the 24-h measured data, and the collected flow, speed and density are scattered, and the traffic parameters are obtained from the expressway. This method has a great dependence on the 24-h measured traffic flow data. However, in the actual data acquisition, the detector can not spread all the basic sections of the whole city highway, even if there is a section of the detector can not get all the traffic conditions of the traffic flow characteristics of the parameters. At this point, the use of curve-based traffic flow feature parameter estimation method will fail or can not get the correct value.

Most of the research on traffic flow characteristic parameters is based on the measured data of highway, and the research on traffic flow characteristics of ring expressway is less involved. In 2007, Guo jifu, Chen dashan and et al. and the ChangJiang Scholar Research Center of Beijing Jiaotong University used the measured data of expressway traffic flow to calibrate the traffic flow parameters. The results show that the characteristics of expressway traffic flow are better than Van Aerde model. In the dynamic simulation software - INTEGRATION, Van Aerde's model (2009) was used to calibrate the traffic flow model parameters. Zhao [9] based on the measured traffic flow data of Beijing expressway, her paper studies the characteristic model of urban expressway traffic flow based on physical attributes, which indicates that the model can better reflect the traffic flow characteristics of urban expressway. Cao et al. [10] based on the measured traffic flow data in Beijing to establish sub-phase flow, speed and density relationship model. Shao et al. [11] studied a new dynamic model through the relationship between microscopic and macroscopic parameters. According to the measured data, the air speed-density, flow rate-density model and flow rate-speed model were established by linear regression analysis. The conclusion shows that the traffic model can reflect the air traffic characteristics. In the paper written by Zhu et al. [12] using mathematical statistics distribution fitting observation data, using Van Aerde's model to study the snow day on the highway traffic flow impact. In the study of free flow speed, domestic scholars use the neural network quantitative and statistical analysis method to carry out the study of free flow speed, and analysis of the adverse weather conditions, The influencing factors of expressway free flow speed and traffic flow characteristics. In the study of traffic capacity, Yaping [13] and Lieqiang and Fu [14] used the capacity analysis method based on mutation analysis and the calculation model of the traffic capacity of the entrance ramp.

In order to meet the requirements of dynamic traffic situation assessment, this paper puts forward the idea of "traffic flow characteristic parameters" based on the physical characteristics of the Ring Expressway, which is based on the existing methods to calibrate the dynamic traffic model of the metropolitan expressway, and a traffic flow characteristic model of Ring Expressway based on physical attribute is established in order to overcome the limitation of characteristic parameter calibration of traffic flow of expressway under favorable data.

## 2 Influencing Factors of Characteristic Parameters of Traffic Flow in Expressway Around

The main factors influencing the traffic flow characteristic parameters of the basic road sections of the road include: road geography, road traffic composition, driver characteristics, road management control strategy. Road geographic features, road traffic composition and traffic data of road-related facilities are regarded as the basic attributes of the road section, called the physical attribute or physical characteristics of the section [15]. In this paper, the concept of 'physical attribute' is put forward for the section of the expressway section, which expresses the basic attributes of the road section. Put forward the 'physical characteristics of factors' to express the physical attributes of the quantization value. Physical characteristics include: geographic characteristics of the property data (such as cross-section width, lane number, slope, entrances and other important facilities, etc.), with the geographical characteristics of the characteristics of the traffic flow data (such as carts import and export flow, etc.). The number of lanes and lane width is the most basic attribute of the Ring Expressway, which has a direct impact on the traffic flow. The carts reflect the traffic composition of the Ring Expressway intuitively. Therefore, the paper analyzes the influence of the number of lanes, lane width, trolley rate and entrances and exits on the characteristic parameters of traffic flow in the expressway section.

- Number of Lanes

The number of lanes as one of the most basic physical characteristics of the section of the Expressway section is the number of unidirectional traffic lanes. The driver's available space increases with the number of lanes, thus affecting the traffic flow characteristics of the Ring Expressway. The number of lanes on the Ring Expressway is 3 lanes or 4 lanes, and the lanes below 3 lanes and 4 lanes are scarce and not representative. Therefore, when the influence of lane number on the traffic flow characteristics is studied, the influencing factors of lane number are {3, 4}.

- Average width of Lanes

In the Road Engineering Technical Standards, the lane is to provide a variety of vehicles vertical arrangement, safe and comfortable through the road belt part. The width of the lane is required for the various vehicles to be driven at different speeds in accordance with the size and speed of the vehicle for driving comfort and traffic safety. Because this paper mainly studies the traffic flow characteristics and traffic flow model of the Expressway, the lane width is taken as the design value of the lane width of the

Expressway section, and the average lane width (ALW) is used as the lane width of the Ring Expressway. The formula is as follows:

$$ALW = \frac{1}{Laneno} \sum_{i=1}^{Laneno} W_i \tag{1}$$

In this formula, $W_i$ is the lane width of the $i$-th ($i$ = 1, 2, …, Laneno) lane of the section. In the ideal state, the basic section of the highway lane width should reach 3.6 m. If the lane width is insufficient, the driver must reduce the speed, widen the distance between the vehicles on the same lane, and maintain a large vertical clearance to compensate for the lack of lateral width. Therefore, the average lane width will have a significant impact on the traffic flow characteristics of the Expressway.

- Big car rate

Heavy Vehicle Ratio (HVR) refers to the ratio of large traffic to all traffic flows on the Ring Expressway. Due to the complexity of the types of vehicles on the Ring Expressway, the performance of large vehicles and the road resources are different from those of small passenger cars. Therefore, the ratio of large vehicles has a significant impact on the traffic flow characteristics of the expressway.

- Import and export related factors

Ring Expressway entrances and exits, the distance is small, traffic complicated. As a result, the traffic flow on the Ring Expressway is greatly influenced by the large number of entanglement. In order to facilitate the calculation, this paper uses a measure of the importance of the import and export of physical quantities, that is, import and export flow (Ramp Volume, shortened form RV) as a ring around the highway import and export factors. The greater the entrances and exits, the higher the frequency of intertwined and lane changes, and the greater negative impact on the traffic flow in the main road of the Ring Expressway.

## 3   Traffic Flow Characteristic and Parameter Model

### 3.1   Model Building Ideas

First, the physical attribute data (physical characteristic parameters: lane number, average lane width, big car rate, etc.) and traffic flow data (flow rate, speed, occupancy rate, etc.) are obtained by using some sections of the Ring Expressway. The traffic characteristics of the traffic flow have a certain influence on the physical characteristics, using Van Aerde traffic flow model to estimate the traffic flow characteristic parameters around the city's highway, and finally the establishment of the Ring Expressway physical characteristics (independent variable $x$) and each traffic flow (Dependent variable y), then estimate the parameters and get the characteristic model of traffic Flow in ring expressway based on physical attributes $y_i = f(x)$ (as shown in Fig. 1).

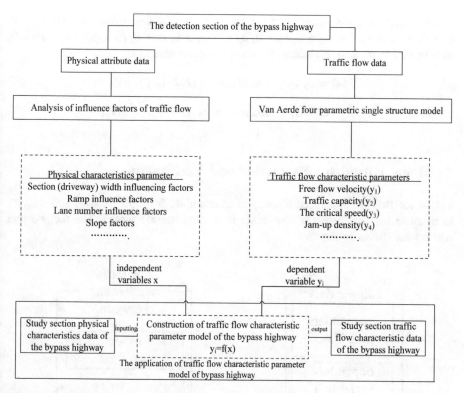

**Fig. 1.** Establishment of characteristic model of traffic flow characteristic of ring expressway based on physical attribute

### 3.2   Characteristic Model of Traffic Flow in Ring Expressway

Guided by the 'expressway traffic flow characteristic parameters into linear function relationship between physical properties and the conclusion', the highway-ring around the preset lane free flow speed (FS), the average capacity (CAP), critical speed (SC), jam-up density (JD) a linear relationship between the physical properties and has the following:

$$FS = a_0 + a_1 \times Laneno + a_2 \times ALW + a_3 \times HVR + a_4 \times RV \tag{2}$$

$$CAP = b_0 + b_1 \times Laneno + b_2 \times ALW + b_3 \times HVR + b_4 \times RV \tag{3}$$

$$SC = c_0 + c_1 \times Laneno + c_2 \times ALW + c_3 \times HVR + c_4 \times RV \tag{4}$$

$$JD = d_0 + d_1 \times Laneno + d_2 \times ALW + d_3 \times HVR + d_4 \times RV \tag{5}$$

In the formula: Laneno, ALW, HVR, RV, respectively, for the number of lanes around the city highway, the average lane width (m), the average car rate (%), import and export flow (pcu/h). The: $a_0$, $a_2$, $a_3$, $a_4$, $b_0$, $b_1$, $b_2$, $b_3$, $b_4$, $c_0$, $c_2$, $c_3$, $c_4$, $d_1$, $d_2$, $d_3$ there are the parameters to be determined.

Qualitative analysis shows that the parameters to be determined: $a_1 > 0$, $a_2 > 0$, $a_3 < 0$, $a_4 < 0$; $b_1 < 0$, $b_2 > 0$, $b_3 < 0$, $b_4 < 0$; $c_2 > 0$, $c_3 < 0$, $c_4 < 0$; $d_1 < 0$, $d_2 > 0$, $d_3 > 0$; $c_1 \approx 0$, $d_4 \approx 0$. Equations (4) and (5) can be written as:

$$SC = c_0 + c_2 \times ALW + c_3 \times HVR + c_4 \times RV \tag{6}$$

$$JD = d_0 + d_1 \times Laneno + d_2 \times ALW + d_3 \times HVR \tag{7}$$

## 4   Model Parameter Estimation and Verification

Taking the Beijing high speed beltway as an example, the paper chooses 130 samples to estimate the parameters of the traffic flow characteristic model. The method and process are shown in Fig. 2.

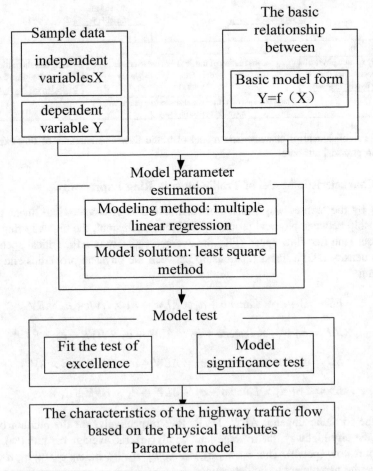

**Fig. 2.**   Parameter estimation, test method and flow of traffic flow characteristic parameter model

The beltway using average method to physical characteristics after the dimensionless processing, with the help of FS, CAP, SC and JD four traffic flow characteristic parameter model by SPSS22.0 software to estimate the model parameters and test, the results are shown in Table 1.

**Table 1.** Model estimation and test results.

| Model | $R$ | $R^2$ | Adjust $R^2$ | F checkout | |
|---|---|---|---|---|---|
| | | | | F | Sig. |
| FS | 0.723 | 0.523 | 0.519 | 7.089 | 0.000 |
| CAP | 0.716 | 0.513 | 0.499 | 10.653 | 0.000 |
| SC | 0.756 | 0.571 | 0.565 | 41.980 | 0.000 |
| JD | 0.705 | 0.497 | 0.489 | 7.350 | 0.000 |

Table 1 shows the results: free flow speed (FS), capacity (CAP), critical speed (SC) and jam-up density (JD) four traffic flow characteristic parameter model of the coefficient of $R^2$ is about 0.5, of which the critical $R^2$ value of the SC model is the highest, and the $R^2$ value of the JD model is the lowest, indicating that the model fit is acceptable. For the model significance test, the F test, the Sig. Values of the four models are 0, the original hypothesis of the F test is rejected, and the four models are significant (Table 2).

**Table 2.** Model coefficients and collinearity statistics.

| Model | Parameter | Estimation | $t$ | Sig. | Collinearity statistics | |
|---|---|---|---|---|---|---|
| | | | | | Tolerance | VIF |
| FS | $a_0$ | 58.174 | 5.372 | 0.000 | | 2.098 |
| | $a_1$ | 1.847 | 1.788 | 0012 | 0.734 | 1.491 |
| | $a_2$ | 2.215 | 2.745 | 0000 | 0.841 | 1.512 |
| | $a_3$ | −1.061 | −2.013 | 0000 | 0.869 | 1.360 |
| | $a_4$ | −1.875 | −2.089 | 0000 | 0.789 | 1.158 |
| CAP | $b_0$ | 1689.461 | 24.587 | 0000 | | 1.413 |
| | $b_1$ | −20.019 | −1.534 | 0002 | 0.708 | 13.148 |
| | $b_2$ | 57.847 | 3.115 | 0000 | 0.027 | 1.215 |
| | $b_3$ | −16.207 | −2.345 | 0000 | 0.028 | 1.176 |
| | $b_4$ | −21.789 | −3.798 | 0000 | 0.809 | 1.365 |
| SC | $c_0$ | 38.478 | 3.578 | 0000 | | 1.112 |
| | $c_2$ | 2.145 | 2.967 | 0000 | 0.901 | 1.017 |
| | $c_3$ | −1.023 | −1.289 | 0034 | 0.971 | 1.123 |
| | $c_4$ | −0.685 | −0.926 | 0056 | 0.864 | 1.250 |
| JD | $d_0$ | 142.525 | 7.738 | 0000 | | 1.189 |
| | $d_1$ | −3.346 | −3.487 | 0000 | 0.734 | 1.178 |
| | $d_2$ | 2.178 | 1.788 | 0000 | 0.857 | 1.019 |
| | $d_3$ | 104.215 | 1.743 | 0084 | 0.980 | 1.243 |

The estimation of the table are consistent with the positive and negative results of the model parameters. The "Sig." Column shows that the original hypothesis of each parameter t test is more likely to be rejected. The "VIF" column shows the variance expansion factor for the estimated values of the parameters used to verify the multicollinearity of the independent variables. Therefore, the parameters in the model estimation have the practical significance of the theory of traffic flow. It can get the traffic flow characteristic parameter model:

- Free flow speed model:

$$FS = 58.174 + 1.847 \times Laneno + 2.215 \times ALW - 1.061 \times HVR - 1.875 \times RV \tag{8}$$

- Capacity Model:

$$CAP = 1689.461 - 20.019 \times Laneno + 57.847 \times ALW - 16.207 \times HVR \\ - 21.789 \times RV \tag{9}$$

- Critical speed model:

$$SC = 38.478 + 2.145 \times ALW - 1.023 \times HVR - 0.685 \times RV \tag{10}$$

- Blocking density model:

$$JD = 142.525 - 3.346 \times Laneno + 2.178\, ALW + 104.215\, HVR \tag{11}$$

## 5 Case Analysis

Xi'an Ring Expressway is the main skeleton hub section of Shanxi Province, which is 88 km long and has 15 interchange bridges. The average of each overpass is about 5 km, which is closed, all interchange and controlled access, two-way and six-lane highway, the schematic diagram shown in Fig. 3.

Based on the GIS map, satellite navigation map and actual survey data, the physical properties of 16 metro expressways were selected to calibrate the traffic flow characteristics of the expressway. The mean value method was used to analyze the physical characteristics of the Xi'an Expressway section as shown in Table 3.

Based on the physical characteristics of the 16 gorges highway sections, the estimated values of FS, CAP, SC and JD for each section are calculated as shown in Table 4. The traffic flow data (flow rate, time average speed, occupancy rate) of the 16 metropolitan highway sections collected by the detector were processed by data processing, 5-min time granularity integration, time-space average speed conversion and occupancy-density transformation, and the FS, CAP, SC and JD of each section of the Expressway are obtained by inputting the VanAerde model with the measured data of the flow rate, as shown in Table 4. The characteristic parameters of the traffic flow

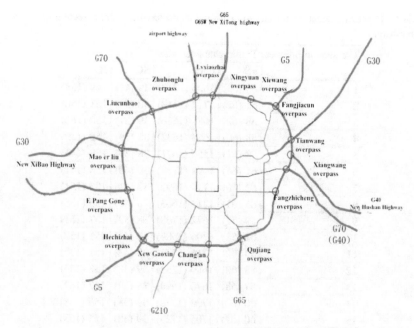

**Fig. 3.** Xi'an Ring Expressway

**Table 3.** Physical characteristics and dimensional characteristic of cross section of Xi'an Ring Expressway.

| The section number | Physical characteristic | | | |
|---|---|---|---|---|
| | Laneno | ALW | HVR | RV |
| 1 | 3 (0.92) | 3.08 (0.97) | 0.07 (0.64) | 390 (0.96) |
| 2 | 3 (0.92) | 3.08 (0.97) | 0.10 (0.91) | 7 (0.02) |
| 3 | 3 (0.92) | 3.08 (0.97) | 0.08 (0.37) | 748 (1.83) |
| 4 | 4 (1.23) | 3.25 (1.02) | 0.15 (1.36) | 908 (2.23) |
| 5 | 4 (1.23) | 3.25 (1.02) | 0.12 (1.09) | 939 (2.30) |
| 6 | 3 (0.92) | 3.17 (1.00) | 0.1 (0.91) | 375 (0.92) |
| 7 | 3 (0.92) | 3.17 (1.00) | 0.1 (0.91) | 0 (0) |
| 8 | 3 (0.92) | 3.17 (1.00) | 0.11 (1.00) | 0 (0) |
| 9 | 3 (0.92) | 3.17 (1.00) | 0.15 (1.36) | 659 (1.62) |
| 10 | 4 (1.23) | 3.25 (1.02) | 0.1 (0.91) | 66 (0.16) |
| 11 | 4 (1.23) | 3.25 (1.02) | 0.1 (0.91) | 547 (1.34) |
| 12 | 3 (0.92) | 3.17 (1.00) | 0.11 (1.00) | 617 (1.51) |
| 13 | 3 (0.92) | 3.17 (1.00) | 0.15 (1.36) | 689 (1.69) |
| 14 | 3 (0.92) | 3.17 (1.00) | 0.13 (1.18) | 178 (0.44) |
| 15 | 3 (0.92) | 3.17 (1.00) | 0.13 (1.18) | 89 (0.22) |
| 16 | 3 (0.92) | 3.17 (1.00) | 0.08 (0.37) | 312 (0.76) |

Note: The data in brackets in the table is dimensionless data for physical features.

**Table 4.** Physical characteristics and dimensional characteristic of cross section of Xi'an Ring Expressway.

| The section number | Traffic characteristics | | | |
|---|---|---|---|---|
| | FS | CAP | SC | JD |
| 1 | 60 (63) | 1696 (1817) | 39 (41) | 149 (158) |
| 2 | 61 (62) | 1712 (1748) | 40 (38) | 153 (156) |
| 3 | 58 (62) | 1681 (1626) | 39 (37) | 150 (163) |
| 4 | 58 (60) | 1653 (1632) | 38 (39) | 156 (150) |
| 5 | 58 (61) | 1556 (1671) | 38 (39) | 153 (146) |
| 6 | 61 (64) | 1694 (1762) | 39 (40) | 153 (151) |
| 7 | 62 (60) | 1714 (1692) | 40 (42) | 153 (154) |
| 8 | 63 (61) | 1713 (1663) | 40 (40) | 153 (165) |
| 9 | 59 (57) | 1672 (1768) | 38 (39) | 157 (161) |
| 10 | 61 (58) | 1706 (1589) | 40 (41) | 151 (152) |
| 11 | 59 (60) | 1716 (1715) | 39 (42) | 151 (170) |
| 12 | 58 (63) | 1680 (1782) | 39 (38) | 153 (154) |
| 13 | 57 (58) | 1670 (1696) | 38 (39) | 157 (162) |
| 14 | 60 (62) | 1700 (1778) | 39 (38) | 155 (168) |
| 15 | 60 (57) | 1705 (1725) | 39 (38) | 155 (163) |
| 16 | 60 (66) | 1700 (1731) | 39 (40) | 150 (158) |

Note: the data in brackets are measured.

**Table 5.** Relative error of estimated value and measured value of traffic characteristic parameters of Xi'an Ring Expressway.

| The section number | Relative error value [%] | | | |
|---|---|---|---|---|
| | FS | CAP | SC | JD |
| 1 | 4.76 | 6.66 | 4.88 | 5.69 |
| 2 | 1.61 | 2.06 | −5.26 | 1.92 |
| 3 | 6.45 | −3.38 | −5.41 | 7.98 |
| 4 | 3.33 | −1.29 | 2.56 | −4.01 |
| 5 | 4.92 | 6.88 | 2.56 | −4.79 |
| 6 | 4.69 | 3.86 | 2.51 | −1.32 |
| 7 | −3.33 | −1.30 | 5.00 | 0.65 |
| 8 | −3.28 | −3.01 | −5.26 | 7.27 |
| 9 | −3.51 | 5.43 | 2.56 | 2.48 |
| 10 | −5.17 | −7.36 | 2.44 | 0.65 |
| 11 | 1.67 | −0.58 | 7.14 | 11.18 |
| 12 | 7.94 | 5.72 | −2.63 | 0.65 |
| 13 | 1.72 | 1.53 | 2.56 | 3.09 |
| 14 | 3.23 | 4.39 | −2.56 | 7.74 |
| 15 | −5.23 | 1.16 | −2.63 | 4.91 |
| 16 | 9.09 | 1.79 | 2.50 | 5.06 |

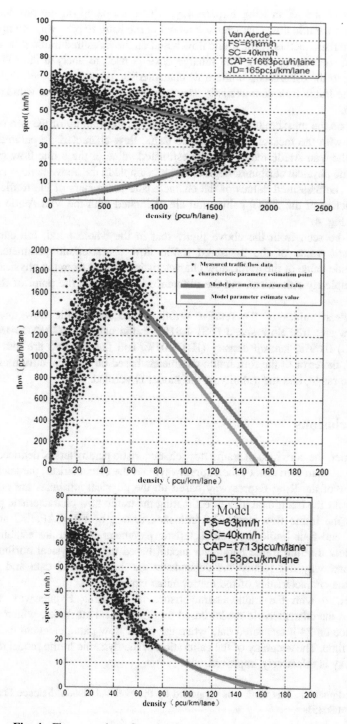

**Fig. 4.** Flow rate chart, flow density chart and density speed chart

parameters of the Xi'an Ring Expressway, which are based on the physical characteristics of the traffic flow characteristic model of the Ring Expressway, compared with
the characteristic factors of the traffic flow based on the measured traffic flow data (flow
rate, speed, occupancy rate). The relative error is used to analyze the fitting effect.
According to the established traffic flow characteristic parameter model, the estimated
value of the traffic flow characteristic parameter and the relative error value are shown
in Table 5.

Based on the measured data obtained by RTMS, the validity of the two methods is
compared with the measured value of the traffic flow characteristic parameters measured by the Van Arede model and the estimated value of the traffic flow parameters
based on the physical attributes. degree. The 8th sample in the above table is taken as an
example. The measured values of the characteristic parameters of the traffic flow and
the parameters of the model calibration are calibrated with the Van Aerde model, as
shown in Fig. 4.

It can be seen from the above figure that in the whole trend, the curve of the
characteristic value of the measured traffic flow characteristic parameter and the
characteristic value of the traffic flow characteristic deduced from the physical attribute
is not completely coincident, but the difference is small, and the point of the trend is
coincide.

The mean values of the relative errors of the four traffic flow characteristic
parameters are: free flow speed (FS) 1.81%, traffic capacity (CAP) 1.41%, critical
speed (SC) 0.69%, jam-up density (JD) 3.07%, and the last two are less than 5%.
Therefore, the error of the four traffic parameters based on physical attributes is within
reasonable range, and the validity of the model is verified.

# 6    Conclusions

In this paper, the core idea of traffic flow characteristic parameters is deduced by using
the physical characteristics of the expressway, and the characteristic parameters of the
traffic flow of the Ring Expressway based on the physical attributes are constructed,
which breaks the traditional idea of estimating the traffic flow characteristic parameters
based on the traffic flow data. The influence factors of FS, CAP, SC and JD are
analyzed, and four traffic flow characteristic parameter models are established. The
results show that the traffic parameter model based on the physical attributes of the
Ring Expressway can reduce the dependence on the measured data and can better
describe the characteristics of the metropolitan expressway.

Combined with the traffic characteristics of the Ring Expressway, this paper
establishes the characteristic model based on the physical attributes, which avoids the
dependence on 24-h measured data when the traffic flow parameters are estimated. At
the same time, The accuracy of the calibration of the dynamic traffic model of the Ring
Expressway is effectively improved.

**Acknowledgments.** This work was supported by the National Natural Science Foundation of
China (51468035).

# References

1. Fang, X.: Research on The Key Technology of Predicting Metropolitan Beltway Highway Traffic Volume. Chang'an Univeisity, Xi'an (2009)
2. Jiang, R., Wu, Q.S.: Study on the complex dynamic properties of traffic flow the micro and macro modeling. J. Graduate Sch. Chin. Acad. Sci. **23**, 848–854 (2006)
3. Greenshields, B.: A study of traffic capacity. High. Res. Board Proc. **14**, 448–477 (1935)
4. Pipes, L.A.: An operational analysis of traffic dynamics. J. Appl. Phys. **24**, 274–281 (1953)
5. Pipes, L.A.: Car-following models and the fundamental diagram of road traffic. Transp. Res. **1**, 21–29 (1967)
6. Wu, D.: Based on Information Fusion Technology of Expressway Traffic Flow Quidance System Model Research. University of Electronic Science and Technology of China, Chengdu (2009)
7. Rakha, H.: Validation of Van Aerde's simplified steadystate car-following and traffic stream model. Transp. Lett. Int. J. Transp. Res. **1**, 227–244 (2009)
8. Hongping, L., Yulong, P., Zhongliang, Y.: The speed and its influencing factors of free flow of expressway. J. Jilin Univ. **27**, 772–776 (2007)
9. Zhao, N.: Based on The Physical Properties of Urban Expressway Traffic Flow Characteristic Parameter Model. Beijing Jiaotong University, Beijing (2010)
10. Cao, Y., Liu, L., Wang, W.: Study on the speed-density model of urban expressway traffic flow. Logist. Technol. **32**(4), 166–169 (2013)
11. Shao, X., Han, Y., Hu, Y.: A new macroscopic traffic flow model. J. Northeast. Univ. **35**(7), 1056–1058 (2014)
12. Zhu, X., Wang, Y., Shao, X.: Characteristics of highway traffic flow in snowy weather. J. Wuhan Univ. Technol. **40**(1), 95–98 (2016)
13. Yaping, Z.: Research on traffic flow characteristics and traffic capacity of the basic section of the expressway. Doctoral Dissertation of Harbin Institute of Technology, Harbin (2005)
14. Lieqiang, X., Fu, W.: Calculation model of road entrance on ramp junction capacity. J. Huazhong Univ. Sci. Technol. **9**, 91–93 (2006)
15. Transportation Research Board: Highway Capacity Manual 2010. Transportation Research Board of the National Academy of Science, Washington (2010)

# Method to Decompose Regional Travel Demand Model - Case Study of Kraków Region

Rafał Kucharski[✉], Tomasz Kulpa, Justyna Mielczarek,
and Arkadiusz Drabicki

Faculty of Civil Engineering, Cracow University of Technology,
Kraków, Poland
{rkucharski, tkulpa, justyna.mielczarek,
adrabicki}@pk.edu.pl

**Abstract.** In this paper we propose a new approach to model regional travel demand with a four-step model. Based on a comprehensive travel survey results for the Małopolska region in Poland (over 3mln inhabitants) we analysed the demand model with specific focus on regional trips. As a result, we introduce a modified model structure where long-distance trips are exposed. We breakdown the demand to four demand strata of various destination types to improve regional demand model quality. Such decomposition allows better representation of regional travel demand as compared to the classical four-step approach.

**Keywords:** Demand model · Regional demand · Four-step model

## 1 Introduction

Majority of surveyed regional travel demand are short-distance trips (over 90% of all surveyed trips). Consequently, an aggregate representation of travel demand may lead to underestimation of long-distance trips in regional demand model. This might occur at the trip generation, trip distribution, mode-choice as well as route-choice stages, where the goodness-of-fit would be preserved for short-distance trips but poorly reproducing long-distance trips, which are often underestimated and inaccurate. In the results section we show consequences of aggregating trips in the regional context.

To overcome this issue we decided to stratify the data, to be able to describe and understand long-distance travel demand at a more disaggregate and detailed level. We assumed that travel behaviour differs significantly with a trip length, i.e. occasional trips to central metropolis differ significantly from daily commuting to the nearest town. We assumed that trips are assigned to one of four different trip destination strata: to the central metropolis (city of Kraków), to main cities (Nowy Sącz and Tarnów), to counties (regional centres) and to other communes. We stratify the trips already at the trip generation level and handle them separately in the demand analysis. Such approach allows a completely independent demand model structures and estimations. Thanks to this, distinct trip purposes, generation rates and mode choice behaviour for each strata

© Springer Nature Switzerland AG 2019
E. Macioszek and G. Sierpiński (Eds.): Directions of Development of Transport
Networks and Traffic Engineering, LNNS 51, pp. 114–124, 2019.
https://doi.org/10.1007/978-3-319-98615-9_10

can be identified. The differences between strata can be exposed and described with appropriate model specification.

Long distance travel stands for a disproportionately large share of person kilometres travelled compared to its share in total generation [1].

Most of demand for mobility is supplied locally, within the nearest area. Travellers minimize their travel costs and tend to commute with lowest disutility, i.e. on shortest distances. This leads to concentration of trip destinations around the origins (homes, workplaces, schools) and pressure to minimize travel distances. Consequently, most of mobility is local and only some (infrequent) cases require to go further. From this perspective the demand for regional trips is strongly and negatively correlated with attractiveness of nearest neighbourhood.

We introduce a method modifying the classical four-step demand model and its estimation from the survey data. Yet we decided to stratify the data at the trip generation level. We presumed that the total trip generation (observed in the survey) is supplied at various destination strata. We assumed that most of the demand can be supplied internally within the same zone, since majority of observed trips were intrazonal. If the demand cannot be supplied at the origin, the further destinations need to be found. Hopefully it can be supplied at the closest commune or the nearest town. Otherwise commuter is forced to travel to one of the main cities or even to the metropolis. In such hierarchical framework the decision maker is willing to supply his needs at the closest destination possible. This incorporates two opposing factors making the model interesting: one is willingness to minimize travel costs and another is limited supply at the destinations (i.e. the demand to go to the cinema cannot be supplied at the destination without the cinema). This is the rationale behind the proposed stratification. We presume the commuters stratify themselves and assign to respective strata based on their demands. Such choice process can be well described with a random utility model. Unfortunately, current underlying process at regional level seems to be hardly observable and random, and estimating it with classical approach from the survey data may lead to significant biases. In this, initial stage we proposed a fixed-share approach, where total generation is distributed among strata with a fixed shares. However, we experimented with more complex models (i.e. nested and multinomial logit) and obtained promising results which, in future, can complement the model proposed in this paper. The major contribution of the paper is the proposed decomposing travel demand method which can substantially improve the demand representation in regional trip assignment model. It allows much more accurate and reliable approach to estimating the travel demand and travel behaviour in long-distance trips.

## 2 Background

Traditionally the focus of travel demand models were urban areas, where travel problems (mainly traffic congestion) are most evident. For this reason analytical, quantitative models were proposed of which the four step model is the most popular [2]. Nowadays, the accurate and complete multi-modal dynamic models of urban transport systems are becoming available. On the other hand the long-distance travel behaviour seems to be less investigated.

The national models exists for most of developed countries, usually designed as some kind of simplification derived from urban four-step demand model. Nationwide models have different purposes, thus their methodology is different from the urban ones. Since the typical trip length in national model is above one hour, instead of peak-hour, the whole day (AADT) is modelled. This raises some issues, because the daily capacity of road network does not have a physical meaning. Also, the trip purposes are different and do not focus on commuting but include also leisure, tourism, family affairs and business trips. The heterogeneity of travellers is more pronounced, since the long-distance mobility strongly depends on the socio-economical groups [3]. At the national level the average daily trip generation is typically used instead of the peak-hour, and the seasonal variability is often taken into account. The data collected to build long-distance demand models is usually collected during the travel survey programs, executed on representative population cyclically (like the one used in the paper). Unfortunately, despite the high number of recorded trips, the long-distance trips are under-represented and the estimated statistical models remain weakly determined.

In this paper we focus on regional models, which is not clearly delimited. In extreme cases, regions of big countries may have the size of smaller countries, for instance California with 16 million inhabitants is modelled with a regional model, while The Netherlands of the same population are divided into regional models. For this reason it is hard to clearly describe the methodology of regional demand modelling, which is somewhere between the agglomeration and national level. To define the regional model authors need to decide on transport network representation (detailed vs. simplified), space discretization (dense vs. sparse) and level of detail at the demand model (trip purposes, times of day, socio-economic groups), etc. In this paper we work with the Małopolska Region (Poland) of over 3 million inhabitants and diameter of ca. 100 km. The notable regional models reported in the literature are Dutch regional models [3], which we reference for details of regional models methodology.

Since there is no solid method to model regional demand, there is number of opened issues. Out of many reported, we refer to: [4] where alternative distribution models are presented and [5] with land-use interactions, where the origin attributes plays a crucial role. The similar approach to the one presented in this paper by [6] where instead of using gravity based trip-distribution the trips are clustered based on demand data. The mathematical framework for stratified distribution model can be found in [7], further complemented with the concept of dominance (implicitly followed in the proposed method) by [8].

## 3    Method

The objective in regional demand models is to reproduce the passenger flows over the regional, multi-modal transportation network. In particular we are interested in the reliable estimate for $q_a$, a number of passengers $q$ travelling along arc $\alpha$ of the network during a given time period (usually the day or peak hour). Passenger flows are obtained from the demand model $DM$ with $x$ being input and $\alpha$ being set of parameters. We detail the demand model below and generically formalize it with:

$$\mathbf{q} = DM(\boldsymbol{\alpha}, \mathbf{x}) \tag{1}$$

We use the bold (vector) notation to denote the network-wide values, e.g. $\mathbf{q} = \{q_a: a \in A\}$. The transportation network is defined as a directed, connected graph $G = (N, A)$ of nodes $n \in N$ and arcs $a \in A$, which can be used by various modes $m \in M$ of transport (walk, bike, car, bus, train, etc.) at respective travel costs $c_{m,a}$. The area is divided into traffic analysis zones, a subset of graph nodes $Z \subseteq N$, where trips can start and end. The demand for trips is then defined for origin destination pairs $q_{od}$: $o, d \in Z$.

In this paper we propose a novel structure of the demand model ($DM$) which is better suited to reproduce regional trips. We will introduce it as a modification to the classic four-step demand model ($FSM$) which we define below.

## 3.1 Classical Four-Step Demand Model

The four steps of $FSM$ which allow consistent reproduction of mobility process are: trip generation ($TG$), trip distribution ($TD$), mode choice ($MC$) and assignment ($TA$). The input data to compute the $FSM$ are the spatial variables defined for origin and destination zones (e.g. population, workplaces, shopping area, etc.), the multimodal, parametrized transportation network (arc costs $c_{m,a}$). $FSM$ can be introduced as a sequence of four steps executed on the input data, with output being network flows $q$. We formalize it with the following sequential computation of flows $q$ in increasing detail:

$$\mathbf{q}_{a,m} = TA(\alpha, \mathbf{q}_{od,p,m} = MC(\alpha, \mathbf{q}_{od,p} = TD(\alpha, \mathbf{q}_{o/d,p} = TG(\alpha, \mathbf{x}_{o,d})))) \tag{2}$$

with respective steps briefly introduced as follows. In the trip generation ($TG$) the production at origins $q_{o,p}$ and attraction at destinations $q_{d,p}$: $o, d \in Z$ are defined for trip purposes $p$. The trip purposes are typically at least Home-Work, Home-Education, Home-Other [9]. The trip generation model is typically estimated using results from the revealed-preference survey. The objective of the estimation is to explain the number of generated and attracted trips with available zone variables $x$ using appropriate formulas. Typically, the simple linear regression is used for production and attraction:

$$q_{o,p} = TG(\alpha, \mathbf{x}_o); q_{d,p} = TG(\alpha, \mathbf{x}_d) \tag{3}$$

with the variables $x_i$ typically being: population, number of workplaces, number of places in schools, area of services, shops, etc. In the paper we will illustrate the method in the next section with the model estimated from the 2012 Małopolska Comprehensive Travel Study.

In the second step, the origin demand is supplied at the destinations in the trip distribution ($TD$) procedure. Typically gravity model is used with trip impedance estimated from the survey data, defined generically:

$$q_{od,p} = TD(\alpha, q_{o/d,p}, C_{od,p}) \tag{4}$$

where:

$q_{od,p}$ - number of trips between origin and destination.

Number of trips between origin and destination is computed per trip purpose $p$ using the origin production $q_{o,p}$, destination attraction $q_{d,p}$ and travel costs $C_{od,p}$. The perceived cost $C$ to travel between origin and destination can be seen as a function of actual (physical) network costs $C_{od}$ (distance, time, comfort, reliability, etc.) and the parameters $\alpha$ allowing to reproduce the actual destination choice behaviour. Usually the cost impedance is expressed as a probability density function of exponential or log-normal distribution shape with trip probability decreasing with increasing travel cost.

When the trip matrices are computed in the trip distribution step, the mode-choice model *(MC)* is applied to estimate modal shares of the considered mode alternatives $M$. The share of trips $\alpha_{od,p,m}$ conducted with mode m to travel in purpose $p$ from origin o to destination d can be expressed with a generic multinomial logit model:

$$p_{od,p,m} = \exp(U_{od,m}) / \sum_{m' \in M} \exp(U_{od,m'}) \tag{5}$$

where:

$U_{od,m}$ - utility of alternative $m$ from the available choice-set $M$.

The utility of using mode m is in general a function of travel costs and constants. In the estimation the shape of utility function is determined to maximize the likelihood to reproduce the actual mode choice revealed in the survey. Mode shares are then used to extract number of trips conducted with a respective mode m from origin o to destination d during peak-hour h on a trip purpose m:

$$\mathbf{q}_{od,m,p} = MC(\alpha, q_{od,p}) = p_{od,m,p} \cdot q_{od,p} \tag{6}$$

The mode specific origin-destination matrices are finally assigned to the paths in respective modes. The flow $q_{od,m,p}$ is distributed to paths $k \in K_{od}$ with a given shares $R_{k,od}$, computed by subject to path costs and path-choice criteria, typically being the travel time, cost or distance. The paths are defined as a sequence of network arcs, so that one can directly obtain then arc flows as a sum over path flows for paths using the arc:

$$q_{a,m} = \sum_{o,d \in Z} \sum_{k \in K_{od}} R_{k,od,m,p} \cdot q_{od,m,p} \cdot i_{k,a} \tag{7}$$

where:

$i_{k,a}$ - is equal one if path $k$ traverses arc $a$ and zero otherwise.

Since usually the assignment is an iterative procedure (like the classical Wardrop equilibrium) without a closed form, we introduce it through the following generic function of demand flows and network mode-specific costs:

$$\mathbf{q}_{m,a} = TA(\boldsymbol{\alpha}, \mathbf{q}_{k,od,mp}) \tag{8}$$

The above formulations (3–8) can be used to compute the FSM when applied sequentially to solve Eq. 2. Each of sub-models exposes different aspect of mobility and can be parametrized to match the mobility behaviour observed in the survey.

### 3.2 Regional Four-Step Demand Model

In this part we elaborate on the above model and its applicability in the regional context. The implicit assumption of the four step model is the population homogeneity. Although the trips are classified per purposes and often the users classes (working/non-working, low/high income) are introduced, the trips in the urban context remain spatially homogeneous. Since the *FSM* was conceived to reproduce urban trips, all the trips, regardless the purpose and user class start and/or end within the city, which implies the specific distance, duration and behaviour in the trip. This no longer holds true for regional model, where significantly different trips are conducted: short urban, long urban, short extra-urban, short intra-urban, long intra-urban, etc. Since most of mobility is short distance, the longer trips are under-represented in the estimation of particular parts of four-step demand model. Namely, the trip productions and attractions are estimated to reproduce mainly local mobility (which is in huge majority) which can result in biased attractiveness where destinations of longer trips are underrated. Moreover, the trip purposes usually differ. Short trips are usually regular trips to work, school, shopping etc., while longer trips are often occasional with purposes which are usually aggregated into "other" or "non-home-based" in urban models (e.g. hospital, opera, leisure, business trips). In the regional context, also the mobility rates may be substantially different and mobility of urban and suburban areas may be substantially higher than in remote, rural areas. This becomes even more evident at the trip distribution level, when long trips are lost in the right-hand tail of distance decay function. The objective function of the estimation is driven by the short term trips and there is not enough degrees of freedom to reproduce the specifics of longer trips. Likewise, the mode-choice behaviour is significantly different for regional trips, which are infrequent and the both factors driving mode choice and their randomness are presumably different from the long-term context.

For the above reasons the classical four-step demand model (Eq. 2) seems to not suitable represent the regional trips. To overcome the above limitations we propose the following modifications aimed to improve the model applicability in regional context. The improvements are based on the results of the 2012 Małopolska Comprehensive Travel Study and the conclusions that were made. The proposed changes may be summarized with the following:

- we extend the trip generation and express it as a function of accessibility,
- we stratify trips per destination type, we introduce the trips strata with destinations: at the central metropolis of the region, at the main cities of the region, at the counties (regional centres), at other communes and internal (within the same zone),
- we propose the choice model where for total trip generation is distributed between the above destination strata,

- we introduce the strata-specific demand models, better reproducing different specifics of the strata.

What we achieve by following changes in the FSM. First, the trip generation becomes a function of potential accessibility [5], expressed as a function of travel costs from/to a given zone and the attractiveness of other zones, generically expressed with:

$$PA_o = \sum_{d \in Z} \exp(-\alpha \cdot c_{od}) \cdot x_d \qquad (9)$$

where:

$\alpha$ - the parameter,

$x$ - the zone characteristic used to express its potential (most likely identical like the one used to estimate trip attraction $q_{d,p}$).

Both of them are estimated to fit the data. The PAo is used as a component in the trip generation, estimated from the data (we exposed positive correlation between number of trips and accessibility in the field data, unfortunately the sample size was too limited to estimate the model).

Each of the surveyed trips can be assigned to one of above strata by looking at the destination zone. Based on this we can estimate the probabilistic choice model where the probability of selecting respective stratum is given. In this paper we assume fixed shares $T_{s,m,o}$ used as an intermediate between trip generation and trip distribution. Alternatively, instead of the fixed shares, we estimated the discrete choice model from the survey data. The results are promising with explanatory variables being accessibility (e.g. number of workplaces) and distance and authors hope to present them in future papers.

The two above are novel elements of the proposed model which allow to consistently stratify the data into separate groups. As we already mentioned the travel behaviour is strictly related to the strata. This refers to the trip generation formulas (Eq. 3), parameters of the trip distribution (Eq. 4), mode utilities in the mode-choice model (Eq. 6) and finally the route-choice parameters (Eq. 8). So that all the formulas of the FSM become now stratum-specific, e.g. the trip generation (Eq. 4) can now be defined for each strata separately allowing for better fit to the actual behaviour:

$$q_{od,p,s} = TD(\alpha, q_{o,p,s}, q_{d,p,s}, C_{od,p,s}) \qquad (10)$$

To summarize we hypothesize that the above formal framework will provide better representation of the current regional travel demand. We formalize those assumptions with the following two hypotheses, that we support with the results in the next section:

- **H1.** Estimated travel behaviour is significantly different in respective strata than the one estimated from aggregated data in at least one of FSM steps of each stratum,
- **H2.** The results of stratified model allow to better reproduce the observed behaviour, especially in the regional context.

## 4 Results and Summary

In this section we illustrate the above method using case-study of Małopolska regional model. We show how strata differ from each other by looking at the resulting trip distribution and time-of-day choice. In the first step we compared trip length distribution for aggregated trips and stratified. As we can see in Fig. 1 the trip length distribution significantly varies between strata. Clear-cut shape for aggregated trips is visible (well-fitting exponential distribution) which becomes more complex for trips to main cities and to metropolis. This suggests that the model estimated for the aggregated trips cannot reproduce the different behaviour in respective strata which supports hypothesis H1.

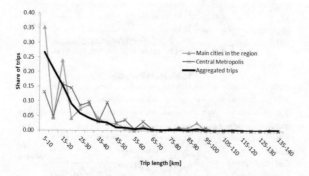

**Fig. 1.** Trip length distribution [km] total for two selected strata

In Fig. 2 trip length distributions disaggregated into respective cities are presented. We can now see similarities and differences and divide the destinations into groups with similar shapes. We noticed two groups: cities with high share of short trips and low share of long trips, and the ones higher share of longer trips. This supports hypothesis H2 by showing that trip distribution differs significantly between strata and cities can be clustered into homogeneous strata.

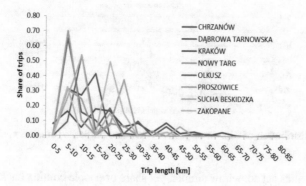

**Fig. 2.** Trip length distribution [km] for counties in Małopolska

Finally, we analysed daily distribution of trips and compared them between the strata. Figure 3 shows number of trips in various purposes executed to respective strata. The central metropolis strongly attracts the morning commute, i.e. work (HBH - blue) and school (HEH - red), with few people coming back from work to metropolis in the afternoon. The trips in other purposes (HOH - green) are mainly attracted in the morning, similarly to the non-home based trips (NHB - violet), yet the afternoon peak for NHB trips is also observed.

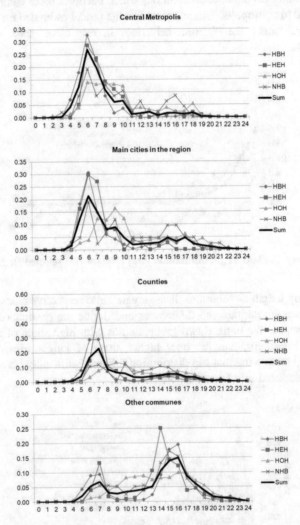

**Fig. 3.** Daily distribution of trips in respective strata and purposes (time on x, share on y axis)

The main cities act somewhat similar, yet share of people coming back from work is visible (around 3 p.m.), which makes afternoon peak higher than for metropolis. In the counties the HOH and NHB trips are uniformly distributed during the day, the peaks are not evident. The morning commuters peak is more pronounced and concentrates around 7 a.m., yet there is a significant number of commuters coming back from work in the afternoon (higher than for metropolis and cities). The other communes are completely different, with more trips attracted in the afternoon. Apart from coming back from school and work, also other purposes and NHB trips are supplied mainly in the afternoon. Such analysis shows that the estimation of peak hour share will provide better results for stratified trips, which supports hypothesis H2.

The above preliminary results serve as a proof-of-concept to propose a dedicated method to reproduce regional trips. We demonstrated that trips shall be stratified as this allows to handle them separately and obtain better fit. We showed in the results that the travel behaviour significantly differs across the strata. Thus, for purposes of regional travel demand modelling we propose to consider above findings and restructure the demand model accordingly. We presume the commuters stratify themselves and assign to respective strata based on their demands. Such choice process can be well described with a random utility model. In this, initial stage we proposed a fixed-share approach, where total generation is distributed among strata with a fixed shares. However, we experimented with more complex models (i.e. nested and multinomial logit) and obtained promising results which, in future, can complement the model proposed in this paper.

This initial results were applied in the Małopolska regional model to propose a novel structure which resulted in model better fitting the actually observed mobility behaviour and of higher descriptive value.

**Acknowledgements.** This paper was developed within R&D project RID2A "Trip modelling and forecasting using different modes of transport" financed by The National Centre for Research and Development and National Roads and Motorway Administration in Poland.

# References

1. Börjesson, M.: Forecasting demand for high speed rail. Transp. Res. Part A: Policy Pract. **70**, 81–92 (2014)
2. Joksimovic, D., van Grol, R.: New generation of Dutch national and regional models-an overview of theory and practice. In: European Transport Conference (2012)
3. He, S.Y., Giuliano, G.: School choice: understanding the trade-off between travel distance and school quality. Transportation **44**, 1–24 (2017)
4. Cascetta, E., Pagliara, F., Papola, A.: Alternative approaches to trip distribution modelling: a retrospective review and suggestions for combining different approaches. Reg. Sci. **86**(4), 597–620 (2007)
5. Stepniak, M., Rosik, P.: Accessibility improvement, territorial cohesion and spillovers: a multidimensional evaluation of two motorway sections in Poland. J. Transp. Geogr. **31**, 154–163 (2013)
6. McNally, M.G.: The four-step model. In: Hensher, D.A., Button, K.J. (eds.) Handbook of Transport Modelling, 2nd edn, pp. 35–53. Emerald Group Publishing Limited (2007)

7. De Grange, L., Ibeas, A., González, F.: A hierarchical gravity model with spatial correlation: mathematical formulation and parameter estimation. Netw. Spatial Econ. **11**(3), 439–463 (2011)
8. Cascetta, E., Papola, A.: A trip distribution model involving spatial and dominance attributes. Comput. - Aided Civil Infrastruct. Eng. **23**(2), 116–124 (2008)
9. Ermagun, A., Fan, Y., Wolfson, J., Adomavicius, G., Das, K.: Real-time trip purpose prediction using online location-based search and discovery services. Transp. Res. Part C. Emerg. Technol. **77**, 96–112 (2017)

# The Concept of Using a Dual-System Tramway in the Upper Silesian Agglomeration

Damian Lach[(✉)]

Faculty of Transport, Silesian University of Technology, Katowice, Poland
damian.lach@polsl.pl

**Abstract.** The article presents the characteristics of the specifications enabling the use of a dual-system tram in the Upper Silesian Agglomeration. Technical and socio-economic specifications have been taken into account. The dual-system tram is a system of public passenger transport connecting the subsystems of rail transport: train and tram. The most widespread model is the "Karlsruhe Model", i.e. the model in which the vehicle is adapted to moving on the tram tracks and railway tracks.

**Keywords:** Transport systems · Public transport · Dual-system tramway
The Karlsruhe Model

## 1 Introduction

Transport systems of areas characterized by a high degree of urbanization require constant improvement and development. The development of urban agglomeration areas and the constant increase in the value of traffic flows determine the need for efficient means of transport in a given area [1]. The origins of urban agglomerations go back to the nineteenth century, when the industrial revolution took place. Less than 70 years after the beginning of the railway era rail transport to city streets was introduced. Along with the demand, the capacity of vehicles grew and transport networks grew. Nowadays, the capacity of trams is maximally utilized and the capacity of tram lines is often exhausted. A suitable solution for the given problem of the capacity of tram lines and vehicle capacity is the connection of two different rail transport systems. The dual-system tram that combines the features of the tram and train can become an alternative urban and agglomeration mode of transport [2]. In the case of using the maximum capacity of tram lines, a good alternative is the underground metro system, which does not interfere with the dense development of city centers. However, the construction of metro systems is cost-intensive and maintaining the entire system requires large financial and organizational expenses. An alternative to metro systems are railway lines running through city centers. In most cases, they are used only for running long-distance and regional (agglomeration) trains, whose stations and stops are located at a considerable distance from each other and do not reach large groups of residents. The economical form can become a dual-system tram, for which the construction of new infrastructure does not require large financial outlays. Stops located on railway lines may be closer to each other and the cost of their construction would be smaller than the

© Springer Nature Switzerland AG 2019
E. Macioszek and G. Sierpiński (Eds.): Directions of Development of Transport
Networks and Traffic Engineering, LNNS 51, pp. 125–131, 2019.
https://doi.org/10.1007/978-3-319-98615-9_11

construction of a standard railway platform or metro station. Such a vehicle is also compatible with the tram network together with the technical infrastructure. Combining all the features of the tram and train there can get an effective means of urban transport, which can solve the problems of tramway lines capacity and meet the requirements of the inhabitants of the agglomeration. The most popular example of a dual-system tram is the "Karlsruhe Model" [3–5]. A model in which trams use the city tram network managed by the city and a railway network managed by the state-owned company Deutsche Bahn. The model assumes the existence of suburban tram lines, which use suburban infrastructure in suburban areas and in urban areas from a tram network. The model may contribute to increasing the attractiveness of public mass transport in suburban areas (agglomeration) by minimizing the number of journeys in which it is necessary to make a change which will reduce the traffic of individual vehicles on city road networks. At present, there are 13 lines of the dual-system tram operating in the Karlsruhe communication region. The article presents a proposal for using of dual-system tramway in Polish environment. Its aim is to search for some possibilities of their use on a selected tram and rail network to develop transport systems in cities and agglomerations.

## 2 Area of Analysis

The Upper Silesian Agglomeration area comprising 19 cities and it is the largest agglomeration of this type in Poland. The population of the agglomeration exceeds 2 million people [6]. The specification of the region forces constant development of transport systems serving both freight and passenger transport [7, 8]. A well-developed tram network in cities and adequate railway infrastructure can become an asset when planning a possible construction of a dual-system tram. The location of cities and considerable distances do not support building a metro system in this area, which could become too expensive. Other options should be considered to help meet the transport needs of residents. The alternative may be railway lines used in the past in passenger trains, on which passenger connections are not organized. In conjunction with the developed tram network there can get an efficiently and effectively functioning system of dual-system tram, which would be integrated with other public transport means. Similarly to the "Karlsruhe Model" in city centers, two-system trams would use the network of tram tracks and in connections between the analyzed cities they would use the railway network managed by PKP PLK (Polish Railway Lines). The result would be a system of quick agglomeration connections improving the transport system of the agglomeration without incurring high costs related to the construction of a new line infrastructure (metro system).

The tram network of the Upper Silesian Agglomeration covers the majority of cities that are part of the agglomeration. The degree of tram network density depends on the specificity of the city. Table 1 shows the cities in the Silesian agglomeration in which there is a tram network.

Out of 19 cities that are part of the Upper Silesian Agglomeration 12, there are tram lines organized by the Communications Municipal Association of the Upper Silesian Industrial District. The most tram lines run through the city of Katowice. This is related

**Table 1.** List of cities in Upper Silesian Agglomeration with tram system.

| City | Number of tram lines | Connections with another cities |
|------|------|------|
| Zabrze | 4 | Bytom, Ruda Śląska |
| Bytom | 8 | Zabrze, Ruda Śląska, Chorzów, Katowice, Świętochłowice |
| Ruda Śląska | 4 | Bytom, Zabrze, Świętochłowice, Chorzów, Katowice |
| Świętochłowice | 4 | Bytom, Chorzów, Ruda Śląska, Katowice |
| Chorzów | 8 | Bytom, Katowice, Ruda Śląska, Świętochłowice |
| Katowice | 14 | Bytom, Chorzów, Świętochłowice, Ruda Śląska, Sosnowiec, Mysłowice, Siemianowice Śląskie |
| Siemianowice Śląskie | 1 | Katowice |
| Mysłowice | 2 | Katowice, Sosnowiec |
| Sosnowiec | 5 | Katowice, Mysłowice, Dąbrowa Górnicza, Będzin |
| Będzin | 4 | Sosnowiec, Dąbrowa Górnicza, Czeladź |
| Czeladź | 1 | Będzin, Dąbrowa Górnicza |
| Dąbrowa Górnicza | 3 | Sosnowiec, Będzin, Czeladź |

to the central character of the city and the structure of the tram network of the agglomeration. In the city of Katowice, tram lines that run to most of the cities covered by the tram network start and finish. A given dependence may be an argument for the creation of a dual-system tram that can improve traffic between cities covered by the tram network and those that are not covered by this system. The number of tram lines in a given city may also be one of the factors affecting the introduction of dual-system tram system to this city, as the number of tram lines in most cases is related to the density of the tram network. Another important factor is the number of inhabitants and functions that a given city has. Organization of the dual-system tram line should also include areas not covered by the ordinary tram system. Such a case is the city of Gliwice, which in 2009 suspended the tram service, replacing it with buses. However, due to the importance of the city on the scale of the agglomeration, when planning a dual-system tram line, the reconstruction or construction of new tram lines should be considered. Figure 1 shows the tram network of the Upper Silesian Agglomeration plotted on the map of the agglomeration.

The scheme includes a tram network on which traffic is routed on regular communication lines. Attention should be paid to the division of the tram network. The western and eastern part of the tram line is connected by only one tram line running in the cities of Katowice and Sosnowiec. Such a state is caused by the historical border of states (Germany and Poland) which resulted in the creation of two separate communication systems in the analyzed area. The connection of the eastern and western part of the tram network of the Upper Silesian Agglomeration can be realized by means of a dual-system tram without the need to build an additional tramway infrastructure for trams. For example, there can analyze the connection of cities such as Bytom and

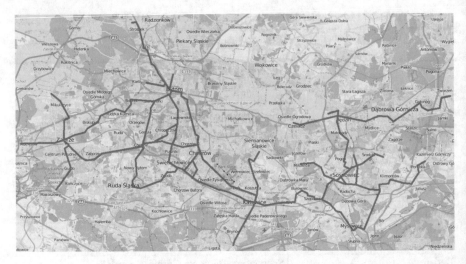

**Fig. 1.** Network of tram system in Upper Silesian Agglomeration (Source: Own using [9])

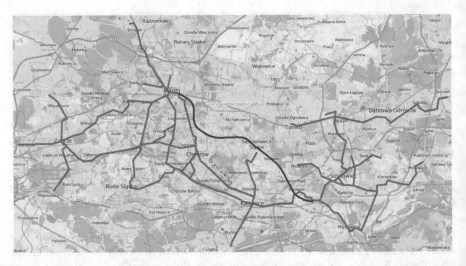

**Fig. 2.** A proposal of a dual-system tram line in Upper Silesian Agglomeration (Source: Own using [9])

Sosnowiec using a two-system tram system. Figure 2 presents an exemplary connection of a dual-system tram running on railway lines no. 161 and 131.

The scheme shows that a dual-system tram would benefit from the city tram infrastructure in Bytom and Sosnowiec. The first link can be located in the Szopienice district in Katowice near Bednorz street where railway sidings excluded from traffic are located. Figure 3 presents the proposal for the location of a rail and tram connection in Katowice.

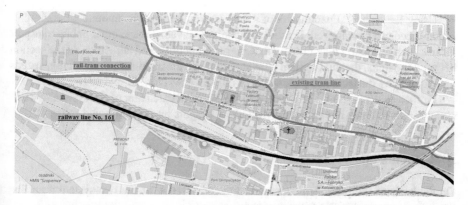

**Fig. 3.** A proposal of a rail-tram connection in Katowice (Source: Own using [9])

The course of the dual-system tram line following the route no. 161 enables the construction of new tram stops. The effect would probably be to increase the attractiveness of public transport in Siemianowice Śląskie and the northern part of Chorzów. The location of given stops should depend on the location of housing estates and the zones in which there are jobs.

The next link would be in Bytom, near Siemianowicka Street. Figure 4 presents the proposed location of the link in Bytom.

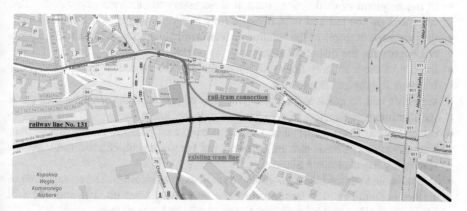

**Fig. 4.** A proposal of a rail-tram connection in Bytom (Source: Own using [9])

The proposed rail-tram connection in Bytom would allow the entry of dual-system trams to the city of Bytom. The effect will be to connect the entire tram network in Bytom with a dual-system tram. Further analyzes should concern interest in a given connection among the inhabitants of the analyzed area and the size of trips made between selected areas along the route of a dual-system tram. The analyzes should also include possible development of the system of dual-system tram for successive sections between the cities of the agglomeration.

Using the technical parameters of a dual-system tram, large amounts of communication speed can be achieved. In relation to individual vehicles and other public transport means, such a solution may turn out to be very competitive and attractive. At the section between the links, the dual-system tram should reach the maximum construction speeds limited only by the technical parameters of railway lines No. 161 and 131.

Further analysis should be devoted to the possible extension of the system of dual-system tram to areas not covered by regular transport by rail (tram and railway).

## 3   Conclusion

On the basis of the presented concept of the construction of the two-system tram line the following conclusions were received:

- the Upper Silesian Agglomeration area, thanks to the increase in urbanization, needs an efficient transport system that will meet the growing requirements of its users,
- achieving maximum values of the capacity of transport network elements dictates the conditions for introducing new solutions,
- building a metro system from scratch would be more expensive than implementing a dual-system tram system,
- a dual-system tram can become an appropriate solution for the problems,
- the presented solution does not require construction or modernization of the entire transport network, which minimizes the costs of its implementation,
- to run a system of dual-system tram, it is necessary to purchase new vehicles that can move on the rail and tram network (this is due to the difference in tensions),
- the presented example of the solution concerns railway line no. 161 and 131 being a rail bypass for the district of Katowice,
- the use of these railway lines will allow the creation of a new transport corridor between Bytom and Sosnowiec,
- the effect will be quick and efficient transport of people in the analyzed area using the maximum parameters of vehicles and railway lines,
- on the railway line no. 161, it is possible to build new platform for the dual-system tram covering the areas that are currently not equipped with efficient rail transport (including Siemianowice Śląskie),
- it is possible to extend the dual-system tram lines to other cities. The construction of further potential railway-tram connections on railway lines no. 131 and 161 will allow for the increase of the impact area of the dual-system tram,
- construction of railway and tramway links should correspond to the parameters of the existing rail and tram network.

# References

1. Macioszek, E., Lach, D.: Analysis of the results of general traffic measurements in the West Pomeranian Voivodeship from 2005 to 2015. Sci. J. Silesian Univ. Technol. Ser. Transp. **97**, 93–104 (2017)
2. Naegeli, L., Weidmann, U., Nash, A.: Checklist for successful application of tram-train systems in Europe. Transp. Res. Rec.: J. Transp. Res. Board. **2275**(1), 39–48 (2012)
3. Kraśkiewicz, C., Oleksiewicz, W.: Dual-tramway in Karlsruhe. Logistics **4**, 4255–4261 (2015)
4. Götz, W.: Zintegrowane Systemy Kolejowo-Tramwajowe w Europie - Stan Obecny i Perspektywy Rozwoju. Technika Transportu Szynowego. **1–2**, 32–36 (2000)
5. Harrasek, A.: Dwusystemowy Tramwaj w Nordhausen. Technika Transportu Szynowego **11** (6), 54–55 (2004)
6. Central Statistical Office. https://stat.gov.pl
7. Macioszek, E., Staniek, M., Sierpiński, G.: Analysis of tends in development of freight transport logistics using the example of Silesian Province (Poland) - a case study. Transp. Res. Procedia **27**, 388–395 (2017)
8. Macioszek, E.: First and last mile delivery - problems and issues. In: Sierpiński, G. (ed.) Advanced Solutions of Transport Systems for Growing Mobility. AISC, vol. 631, pp. 147–154. Springer, Cham (2018)
9. Open Street Map Website. https://www.openstreetmap.org

# Methods of Calming Urban Traffic as an Element of Sustainable Transport Development

Maciej Bieńczak and Szymon Fierek[✉]

Faculty of Working Machines and Transport, Poznan University of Technology,
Poznań, Poland
{maciej.bienczak, szymon.fierek}@put.poznan.pl

**Abstract.** The article focuses on issues related to traffic calming in historic urban areas. The paper presents the changes proposed in Poznań which were subsequently implemented in the PTV VISUM simulation environment. Consequently, conclusions were drawn concerning traffic timetables and modal split. The project was completed with a set of additional recommendations.

**Keywords:** Traffic calming · Traffic simulations · Urban mobility

## 1 Introduction

Whilst shaping sustainable transport systems one needs to consider the following aspects [1]:

- environmental - the transport system should generate as little negative an impact as possible on the environment,
- economic - the transport system should be as efficient as possible at the lowest possible cost,
- social - the transport system should be of the utmost use to society and respond to the reported mobility needs.

The implementation of sustainable transport systems in urban agglomerations is changing from a traditional management approach (including transport planning) focusing on transport modes and infrastructure and on increasing traffic capacity and flow to a holistic approach that takes into account the quality of city life and the expectations of different stakeholders, such as those city dwellers with reduced mobility or businesses operating in the city centre.

Sustainable transport systems can be implemented on the following conditions:

- involvement of stakeholders, primarily the residents, in the decision-making process,
- taking into account all modes and forms of transport, while encouraging a shift towards more sustainable modes of transport (e.g. public transport, non-motorised transport),
- long-term planning of the development of transport,

© Springer Nature Switzerland AG 2019
E. Macioszek and G. Sierpiński (Eds.): Directions of Development of Transport
Networks and Traffic Engineering, LNNS 51, pp. 132–141, 2019.
https://doi.org/10.1007/978-3-319-98615-9_12

- evaluation of the existing and future status using performance indicators,
- regular monitoring to assess the implementation and to make possible adjustments, as well as to share the results of the evaluation with urban users,
- considering external costs for all types of urban transport.

Sustainable urban mobility calls for achieving the following goals [2]:

- ensuring the accessibility of the transport system to all users of urban areas,
- improving traffic safety,
- reducing the adverse effects of urban transport, i.e. reducing $CO_2$ and other,
- pollutants emissions caused by transport, reducing noise as well as excessive and inefficient energy consumption,
- improving the efficiency and cost effectiveness in the transport of goods and passengers,
- ensuring a positive impact on the attractiveness and quality of the urban environment.

One of the results of the introduction of sustainable transport systems is calming the traffic in the city, ensuring greater safety for all traffic participants and better quality of life for city dwellers. Such systems involve shaping the road environment through planning and engineering measures to achieve a comprehensive effect of improving road safety, reducing transport inconvenience and bettering public spaces in built-up areas [3].

Traffic calming aims primarily at improving traffic safety and reducing traffic congestions (mainly transit traffic), but it also aims at revitalising and upgrading urban areas, improving the environment and increasing the attractiveness of public space for its residents and other users.

The means of calming the traffic may be of organizational and infrastructural nature. Safety is ensured mainly by limiting the speed of vehicles through the appropriate shaping of road geometry, street design and traffic organisation elements that physically prevent excessive speeding and other unsafe behaviour of drivers, such as overtaking in the areas where it is forbidden [3].

Most frequently used tools for calming the traffic include:

- delimitation of zones with varied accessibility for car traffic,
- sectional speed limits,
- restricted speed areas with or without priority for pedestrian traffic,
- the introduction of shared space, i.e. areas where all users have the same rights, and vehicles are subject to the 'priority to the right' system,
- replacing junctions with roundabouts or introducing the principle of non-priority roads,
- narrowing of roads (point or sectional) and limiting their capacity,
- physical speed limits imposed by:

  1. speed bumps,
  2. elevated crossings or pedestrian crossings,
  3. altering the axis of the road by introducing traffic islands,

4. altering the axis of the road with alternating narrowing to one lane, e.g. placing small architectural elements on the roadway (e.g. pots with greenery),

- charging fees for driving in selected areas of the city.

The use of these tools has been widely described in the literature, mainly in publications in the field of traffic engineering and urban planning [4–12]. As research shows, such investments contribute to a 15% to 40% speed reduction and they may also reduce the risk of accidents by 12% to 45% [13]. At the same time, they contribute to the reduction of noise and local emission of $CO_2$ and substances harmful to human health.

The views on traffic calming have been evolving and changing with time. There are three approaches to the issue [14]:

- taking the measures consisting exclusively in the limitation of vehicle speed and traffic safety (1980s and 1990s),
- taking the measures to reduce vehicle speed and improve traffic safety, with an emphasis on improving the quality of street space (at the turn of the 20th and 21st centuries),
- introducing transport and urban solutions in which the space influences the culture of mobility (the behaviour on the road, choice of transport means, change of preferences as to the place of residence). This is a view that should prevail in the future and constitutes an extension of the above approach.

The latter two approaches lead to the coexistence of different road users and the creation of high quality public spaces [14].

## 2  Situation in Poznań and the Aim of Research

In Poznań, the work on calming down the traffic was undertaken in 2013. In July 2013, Zone 30 was introduced in the city centre which meant that vehicles are currently allowed to go no faster than 30 km/h. This rule has increased road safety and reduced the speed difference between cyclists and cars. Such solutions are often used in historic parts of cities, where the streets are narrow, the traffic is congested and transit journeys need to be eliminated.

The introduction of Zone 30 enforces several changes to be implemented in numerous places by [15]:

- narrowing of street entrances at intersections,
- structuring parking spaces,
- separating cycling paths and introducing contraflow lanes for cycling,
- switching off the traffic lights,
- introducing the elements of small architecture and greenery,
- junction elevation,
- dedicating paid parking spaces to various suppliers,
- marking new parking spaces for the disabled,
- installation of mats for visually impaired persons.

Thanks to the measures taken, pedestrians as well as cyclists and public transport are given priority. Initially, Zone 30 in the centre of Poznań covered the area between the following streets: Solna Street, Marcinkowskiego Avenue, Święty Marcin Avenue and Niepodległości Avenue. In 2016, changes were introduced on Święty Marcin Street, namely its section from Niepodległości Avenue to Marcinkowskiego Avenue, on Ratajczaka Street, the section from Ogrodowa Street to 27 Grudnia Street, and on Wolności Square. In 2017, as part of the second stage of introducing traffic calming, the zone was extended to include the north-eastern quarter between following streets: Garbary, Podgórna and Marcinkowskiego Ave, Wolnica and Małe Garbary. Ultimately, Zone 30 will be extended and will be in force within the first transport framework (with the exception of the main streets). The borders will be defined by the following streets: Pułaskiego, Roosevelta (in the west), Matyi, Królowej Jadwigi (in the south), Jana Pawła II (in the east), and in the north by the railway tracks and Przepadek Street.

The described changes in the centre of Poznań caused an interest in calming down traffic in the neighbouring districts - Jeżyce, Łazarz and Wilda, which manifested itself in the development of multi-variant concepts of traffic organization, including the introduction of traffic calming solutions. The location of these districts is shown in Fig. 1.

Besides calming the traffic, the concept also assumed:

- transferring transit journeys through the district to the city's main communication system (the so-called communication framework),
- introducing a paid parking zone in Łazarz and Wilda (already in force in Jeżyce),
- maintaining as many parking spaces as possible while facilitating pedestrian traffic.

Despite similar assumptions, the concepts differed in methodology and scope. For instance, only the concept for Wilda included taking an inventory of its parking spaces. All concepts were subject to public consultation, in which the option which was most preferred by the inhabitants was chosen. In general, the options introducing one-way traffic exclusively on local streets were preferred.

The situation described above is a reason to check how the variants chosen by the inhabitants (analysed in the concepts from the point of view of individual settlements) fit into the transport system of the city. This will constitute a stage which may be extended to the whole city (or parts of the agglomeration) in the future. For the purpose of this work, proposals were prepared for changes in traffic organisation (in particular the layout of one-way streets) in Jeżyce, Wilda and Łazarz. Then, changes were introduced in the traffic model of the Poznań agglomeration that had been implemented in the PTV Visum environment. The simulations were followed by conclusions.

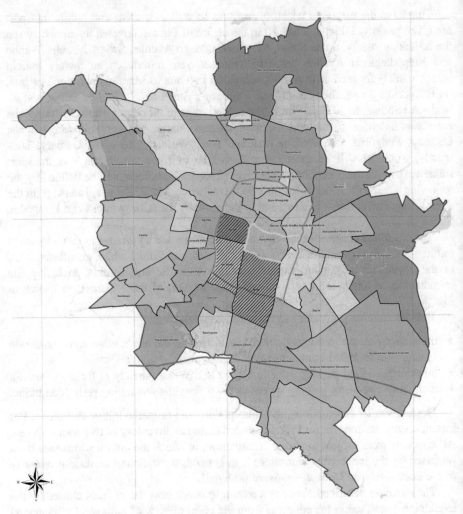

**Fig. 1.**  Auxillary units for the city of Poznań (Source: [16])

## 3   Proposals of Changes for Each of the Districts

Each of the analysed districts is characterized mainly by its frontage construction and urban layout, which has an impact on the proposed solutions of one-way traffic. All the proposed changes are illustrated in Fig. 2.

The district with the most changes is Wilda (11 one-way streets). The decision on the direction of traffic was based on the natural inclination of the area towards the Warta River. Moreover, Wilda features a very irregular road network, which limits the introduction of collision-free one-way street crossings.

In the Łazarz district, a road system based on collision-free crossroads was proposed in its eastern part (with minor deviations due to the need to provide access to the

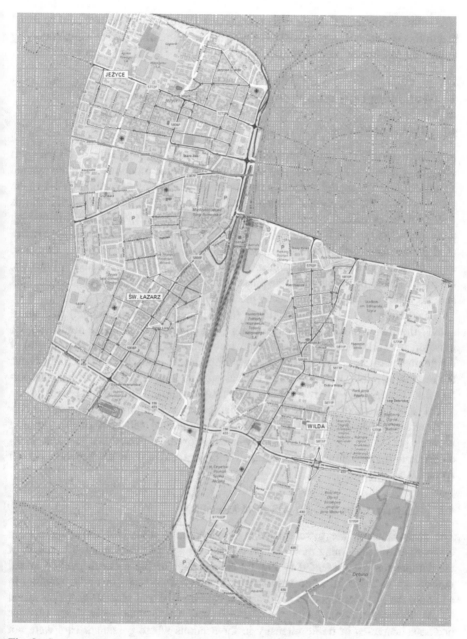

**Fig. 2.** Suggested changes in traffic organization in Jeżyce, Łazarz and Wilda in Poznań (suggested changes in traffic direction are marked in yellow)

properties). In the western part of Łazarz, square shaped quarters have been marked out, rendering it possible to implement a collision-free layout of the streets. However, the strong opposition and misunderstanding of this solution has led to the abandonment of its consideration in this work.

Definitely the least changes were proposed in Jeżyce. However, it should be noted that in this district it was proposed to change the direction of streets on which one-way traffic had previously been in force.

# 4 Modelling and Analyses of Suggested Changes

## 4.1 Travel Modelling

In order to identify the size of traffic flows on particular sections of the city's transport network, a macro-scale analysis was carried out, using the travel model of the Poznań agglomeration. The model contains a mathematical description of relations between the components of the supply and demand structure of transport and its environment, as well as a set of procedures (algorithms) allowing to solve specific decision-making problems related to the functioning of transport systems. One of the basic approaches used in the construction of travel models is the four-stage approach. As the name suggests, it comprises four stages/phases, described by separate mathematical models, i.e. the phases of trip generation, the phase of trip distribution, the phase of modal split and the phase of traffic assignment to the transport network [17].

In addition to the first three stages (underlying the demand model) mentioned above, the travel model must include a supply model, i.e. a model describing the transport network with as accurate a representation as possible (where appropriate) of the geometry of the transport agreement, with different road classes, signalling plans, the routing of public transport lines, etc., as well as a model describing the transport network. Confronting the two models (supply and demand) in the final stage of the model with the use of appropriate traffic distribution algorithms for the transport network allows to determine the traffic flows on individual cells/arcs of the transport network [18]. It is one of the most advanced algorithmic procedures in traffic modelling. It is impossible to model the distribution of traffic over the network without specialized software.

## 4.2 Analysis of the Results of Travel Modelling

The analysis using the travel modelling methodology described above included a comparison of the calming modification variant with the existing variant. The results of the comparison are presented in Fig. 3. The sections of the network marked in red indicate an increase in traffic volume and in green a decrease in traffic volume. In addition, the grey parallel diagonal lines indicate the first and second inner ring roads.

As a result of simulation analyses, it was found that apart from obvious decreases/increases in traffic intensity in the sections where passenger cars were not permitted/permitted to drive, the following changes took place:

- transferring part of the traffic to the 2nd inner ring road, in particular to its southern and western sections,
- increase in traffic volume on Droga Dębińska Street (the street constituting a connection between the city centre and the A2 motorway),

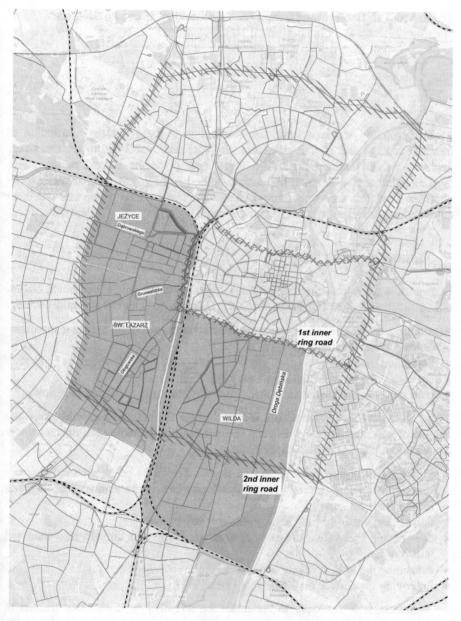

**Fig. 3.** Differences in traffic intensity in different sections of transport network (decreased traffic marked in green and increased traffic marked in red)

- the decrease in traffic volumes in the northern and southern sections of the first inner ring road,
- an increase in the traffic volume in the western section of the first inner ring road.

**Table 1.** Distribution of transport tasks between public and private transport.

|  | W1 | | W0 | |
|---|---|---|---|---|
| Total | 1 205 103 | | 1 205 103 | |
| Public | 434 523 | 36.06% | 431 090 | 35.77% |
| Private | 770 579 | 63.94% | 774 013 | 64.23% |

The proposed solutions have also resulted in a change in the distribution of transport tasks (Table 1). The simulations indicate that over 3 000 trips/journeys are made by public transport on a 24-h basis. Perhaps in the context of the whole agglomeration area it is not a big change, after less than 0.5%, but it needs to be considered that it is done only through changes related to traffic organization, without any interference in the layout of public transport lines.

## 5 Summary

In general, when introducing the solutions described in the article, we should continuously monitor the four fundamental goals that should be achieved by traffic calming:

- restoring the original multi-functional character of streets by reducing the dominance of vehicle traffic,
- improving traffic safety,
- increasing the aesthetic quality of the city,
- improving environmental conditions, with particular emphasis on traffic flow.

As these objectives cannot be achieved through the calming measures alone, the number of passenger cars and delivery vehicles needs to be reduced in the area under consideration. Such activities might include:

- increasing the speed limit for public transport vehicles despite the speed limits regulations binding private vehicles, developing and implementing urban logistics for the analysed area,
- introducing a paid parking zone within the entire area,
- constructing Park&Ride car parks at tram terminuses and strengthening tram and bus transport,
- expanding the system of city bike stations,
- improving the surface on the bicycle lanes.

Moreover, it is necessary to undertake the so-called "soft activities" related to communication with residents and other interested parties. In order to ensure the success of the implementation of such projects, it is necessary to:

- conduct all activities in an open and transparent manner, informing and publicizing them adequately,
- focus on presenting the goal to which the changes are to achieve,
- indicate precisely to what extent any modifications to the project are possible,

- accurately present the traffic calming equipment, including its location and mode of use,
- prepare the methods of effective monitoring of the implemented solutions, before the stage of design is commenced,
- get feedback from all project stakeholders,
- involve the local community at the stage of designing, implementing and monitoring,
- inform the local community about the developments in the process of introducing changes.

# References

1. European Commission: Directorate-General for Mobility and Transport. White Paper on Transport: Roadmap to a Single European Transport Area: Towards a Competitive and Resource-Efficient Transport System. Publications Office of the European Union, Luxemburg (2011)
2. European Commission: Guidelines - Developing and Implementing a Sustainable Urban Mobility Plan. European Commission, Brussels (2013)
3. Slinn, M., Matthews, P., Guest, P.: Traffic Engineering Design. CRC Press, Boca Raton (2006)
4. Harvey, T.: A Review of Current Traffic Calming Techniques. Institute for Transport Studies, University of Leeds, Leeds (1990)
5. Elvik, R.: Area-wide urban traffic calming schemes: a meta analysis of safety effects. Accid. Anal. Prev. **33**, 327–336 (2001)
6. Ewing, R.: Traffic Calming: State of the Practice. Institute of Transport Engineers, Washington (1999)
7. Ewing, R.: Impacts of traffic calming. Transp. Q. **55**, 33–46 (2001)
8. Holzinger, K., Knill, C., Sommerer, T.: Environmental policy convergence: the impact of international harmonization, transnational communication and regulatory competition. Int. Org. **62**, 553–587 (2008)
9. Ewing, R., Brown, S.: US Traffic Calming Manual. American Planning Association, Chicago (2009)
10. Kahn, R., Goedecke, A.K.: Roadway striping as a traffic calming option. Inst. Transp. Eng. J. **81**, 30–37 (2011)
11. Herrstedt, L., Kjemtrup, K., Borges, P., Andersen, P.: An Improved Traffic Environment - A Catalogue of Ideas. Danish Road Institute, Herlev (1993)
12. Zalewski, A.: Traffic Calming as a Urban Planning Problem. Wydawnictwo Politechniki Łódzkiej, Łódź (2011)
13. Proost, S., Westin, J.: Race to the top in traffic calming. Pap. Reg. Sci. **96**(2), 401–422 (2007)
14. Beim, M.: Shared space - evaluation of the idea of space's renewal in German cities. Przegląd Komunikacyjny **11–12**, 10–23 (2011)
15. Poznan City Road Authority Information. http://zdm.poznan.pl
16. Geopoz. https://www.geopoz.pl/portal/index.php?t=20
17. Hensher, D.A., Button, K.J.: Handbook of Transport Modelling. Elsevier, Oxford (2000)
18. Ortuzar, J., Willumsen, L.G.: Modelling Transport. Wiley, New York (2001)

# Traffic Speed Zones and Electric Buses

Krzysztof Krawiec[(✉)]

Faculty of Transport, Silesian University of Technology, Katowice, Poland
krzysztof.krawiec@polsl.pl

**Abstract.** Wide-scale implementation of electric buses imply the necessity of making decisions concerning assignment of electric buses to serve specific bus lines. Ecological issues can be one of the criteria of electric buses allocation to the transport tasks. Due to the fact that electric buses do not emit harmful substances in the place of their use, it is possible to analyze the route of the line in terms the traffic speed zones through which the route passes. The article states that the avoided emission of pollutants emitted by a conventional bus can be measured by the indicator of stops of buses in the traffic speed zones in relation to the number of kilometers driven. This paper presents the methodology to determine traffic speed zones available for city bus, the formulas to determine this value and the exemplary values calculated on the basis of measurements conducted in Polish city Jaworzno.

**Keywords:** Electric buses · Energy consumption · Electromobility
Public transport

## 1  Introduction

Electric buses are starting to play a leading role in low-carbon technologies in public transport. These vehicles perform very well under urban conditions due to the fact that electric motors have higher mechanical efficiency than internal combustion engines. Another advantage of this propulsion is the fact of having full power and maximum torque from zero speed, which distinguishes it from internal combustion engines that require the use of a gearbox [1, 2]. These advantages are particularly visible when driving in urban conditions, which are characterized by frequent repetition of the stopping, starting and braking phases of the vehicle. Unlike conventional Diesel buses, electric ones do not emit harmful substances directly into the atmosphere at their place of use [3]. Replacing the fleet of conventional vehicles to electric ones will contribute to reduce air pollution in urban agglomerations, in particular in city centers. Due to [4], it will be easier to introduce electric vehicles in public transport, comparing to the private cars.

There are no doubts that the process of replacing conventionally-fueled Diesel buses to electric ones is inevitable. In [5], various variants and scenarios for electric buses induction in public transport are presented. Regardless of the strategy chosen, a question should be asked: which lines should be operated by newly purchased electric buses? Answers to this question should consider as economic, ecological, operational and technical issues. Ecological issues are particularly important in polluted areas. One

© Springer Nature Switzerland AG 2019
E. Macioszek and G. Sierpiński (Eds.): Directions of Development of Transport
Networks and Traffic Engineering, LNNS 51, pp. 142–150, 2019.
https://doi.org/10.1007/978-3-319-98615-9_13

should thus think about how to achieve the maximum ecological effect when using the same number of newly purchased vehicles. For this purpose, it is necessary to analyze the route of the bus line. It is particularly advantageous if the routes of bus lines served by electric buses run through city centers and other densely populated areas. Hence, it is necessary to create a methodology that combines the fact of energy consumption (and resulting emissions) with the area through which the line runs.

## 2  Traffic Speed Zones

The area of cities can be divided into several traffic speed zones. Respective traffic speed zones vary among each other by speed limits, accessibility for pedestrians or various groups of vehicles (bicycle, light-duty or public transport vehicles). The main idea of these zones is to provide the required level of road safety. Studies concerning road safety in speed zones are presented, among others, in [6–11].

The rules for determining traffic speed zones vary considerably depending on the country, due to the different nature of road traffic. An exemplary framework for speed-zone guidelines can be found in [13]. In Table 1, Polish framework for determining traffic speed zones is presented. As each of the traffic speed zones listed, can be characterized by different traffic parameters, for example traffic intensity and density or dynamics, the research on driving style and traffic measures-influence on vehicle emissions and fuel consumption has been conducted [14]. Traffic in speed zones behaves differently in city centers and in suburban areas. What is more, there are significant fluctuations in traffic condition depending on the time of day in traffic speed zones. These fluctuations also differ depending on the location of the traffic speed zone.

Furthermore, only three out of the six traffic speed zones listed in Table 1 are adapted for the movement of city buses. City buses are not allowed to move in the pedestrian traffic zone, and also the traffic of city buses is forbidden in the zone of integrated pedestrian, bicycle and car traffic. The practice of routing bus lines indicates that it is avoided to design bus lines in traffic calmed zones. City buses mainly run in moderated speed zones, but often also in heighten speed zones. Occasionally, especially outside city centers, bus lines also run through the high speed zones.

Hence, to study the movement of city buses in the traffic speed zones, changes are to be made to divide the zones presented in Table 1 in such a way so that to take into account only those zones which are affected by city bus traffic. Such analysis should also consider the variations in traffic volume depending on time intervals.

Hence, to study the movement of city buses in the traffic speed zones, city area can be divided into just three of them:

- moderate speed zones in central area of the city (speed limit up to 50 km/h),
- moderate speed zones in suburban area of the city (speed limit up to 50 km/h),
- the areas of increased speed (speed limit equal and above 60 km/h).

The determination of traffic speed zones for the bus traffic analysis may be conducted by following the same rules as for aggregating the area for modelling purposes. The zones - whenever possible - should be compatible with administrative divisions and as homogenous as possible in their land use and population composition. Zones do

**Table 1.** Characteristics of traffic speed zones (Source: based on [12]).

| Name | Speed limit | Description |
|---|---|---|
| Pedestrian traffic zone | N/A | The zone is intended exclusively for pedestrian and bicycle traffic. No access for motor vehicles |
| Zone of integrated pedestrian, bicycle and car traffic | $V \leq 20$ km/h | The area is intended mainly for pedestrian and bicycle traffic. Car traffic access is possible only at very low speed, and is also limited to special vehicles and specific hours of the day |
| Traffic calmed zone | $V \leq 30$ km/h | Area intended for pedestrian, bicycle and car traffic. Vehicle access only at limited speed |
| Moderate speed zone | $V = 40-50$ km/h | Area for car, pedestrian and bicycle traffic. Pedestrian and cyclists can cross traffic flows, even without the use of traffic lights, if only traffic flows are low |
| Heighten speed zone | $V = 60-80$ km/h | The area is mainly intended for the traffic of motor vehicles. Pedestrian and cyclists can cross traffic flows only by using traffic lights or on other levels (footbridge, underpass) |
| High speed zone | $V > 80$ km/h | The area is intended for car traffic only. Pedestrians, cyclists and slow-moving vehicles do not have access to it |

not have to be of equal size - central areas of the city usually they will usually occupy a smaller area than suburban ones. The general rule is that the more congested area is, the smaller zones we make. A full list of zoning criteria is presented in [15]. It should be noted that the determination of borders of traffic speed zones is a very complex issue. In this case there are no rigid rules, each time the process of zoning should be done individually with great care basing on good knowledge of analyzed area.

## 3  An Indicator of Stops of Buses in the Traffic Speed Zones as a Measure of Traffic Conditions

For buses equipped with a diesel engine, the most adverse traffic conditions are urban conditions characterized by frequent braking and leaving the place. However, for electric buses, the fact of frequent stopping of the bus is not a problem due to the fact that the environmental benefits of using electric buses are the most visible in central city areas, characterized by low degree of freedom of movement, high number and high frequency of stops of vehicles and low traffic flow speed. In these conditions conventional buses produce a huge amount of pollutants. What is worse, these pollutants are emitted in the most populated area of the city as this is where the most unfavorable traffic conditions are.

Thus, from the point of view of the allocation of electric buses to public transport lines, it should be borne in mind that the largest possible part of the route should pass through the central areas of the cities. In this situation, we have the greatest environmental benefit understood as non-emitted harmful substances by a conventional bus.

In [16], a number of scheduled bus stops due to the timetable and a number of stops not covered by the timetable are listed as parameters related to the line route that can influence the allocation of electric buses to the transport tasks. Therefore, to assess the avoided emission of pollutants emitted by a conventional bus, a new indicator can be used: the indicator of stops of buses in the traffic speed zones $w_{Z_j}$ in relation to the number of kilometers driven. It should be noted that only these stops of the bus are taken into account that do not result from the timetable. In other words, stops on the bus stops are excluded.

In subsequent traffic speed zones, the value $w_{Z_j}$ takes the following form:

$$w_{Z_j} = \begin{cases} w_{Z_1} & \begin{array}{l}\text{in moderate speed zones in central area} \\ \text{of the city (speed limit up to 50 } \frac{km}{h})\end{array} \\ w_{Z_2} & \begin{array}{l}\text{in moderate speed zones in suburban area} \\ \text{of the city (speed limit up to 50 } \frac{km}{h})\end{array} \\ w_{Z_3} & \begin{array}{l}\text{in the areas of increased speed of the city} \\ \text{(speed equal and above 60 } \frac{km}{h})\end{array} \end{cases} \tag{1}$$

Traffic conditions in the above mentioned zones vary depending on the daytime. There is no doubt that depending on the location of the traffic speed zones within the city, traffic will vary significantly especially during the hours of morning and afternoon peak. It is possible to divide the day into $n$ time intervals.

The formula to calculate $w_{Z_j}$ is as follows for three traffic speed zones:

$$w_{z_j} = w_{z_j}^{tp_i} = \frac{\sum SoB_j}{l_j} \text{for} \quad i = 1, \ldots, n; j = 1, 2, 3 \tag{2}$$

where:
$SoB_j$    - measured number of stops of the bus in $j$-th traffic speed zone,
$l_j$       - total number of kilometers driven during the field studies in $j$-th traffic speed zone,
$w_{Z_j}^{tp_i}$   - value of the indicator of stops of buses in $j$-th traffic speed zone.

It is worth noting that the number of time intervals is not given - it should depend on local traffic conditions. The more time intervals are taken into consideration, the greater is the accuracy. Too detailed division may in turn be unnecessary from the point of view of the essence of the problem.

Calculation of the indicator of stops of buses in the traffic speed zones $w_Z$ in relation to the number of kilometers driven can contribute to the study of the line's ability to be operated by electric buses from an ecological point of view. The higher is the value of $w_Z$ the more the bus line is to be operated electric buses.

# 4  Field Studies in the Area Operated by PKM Jaworzno

The research on the number of stops of city buses were carried out in a medium-sized Polish city of Jaworzno. The field studies were taken in city buses operated by urban public company 'PKM Jaworzno' - an operator of public transport in this area that operates electric buses since 2012. Currently the company has 23 standard and articulated electric buses in operation, which is about a third of the entire bus fleet.

Bus lines operated by PKM Jaworzno are characterized by great diversity in geographical and functional terms. Bus lines have a radiating system with a centrally located stop named 'Centrum'. Both urban and inter-city lines can be distinguished in this transport system. The routes of the bus lines operated by PKM Jaworzno run through densely populated areas, as well as by those which are less populated.

The division into the traffic zones was conducted according to zoning criteria which were discussed above, in particular on the basis of the functional approach to the areas investigated. The division into three traffic speed zones is presented in Fig. 1.

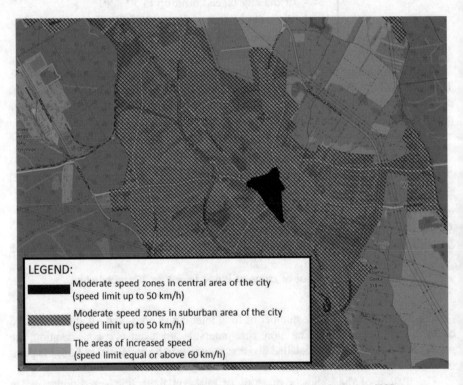

**Fig. 1.** Traffic speed zones in Jaworzno (Source: own basen on [18])

Basing on Rudnicki [17], the following four time periods were adopted ($n = 4$) for the research:

- $05^{00}$–$09^{00}$,
- $09^{00}$–$13^{00}$,
- $13^{00}$–$17^{00}$,
- $17^{00}$–$21^{00}$.

The total amount of 1075 km were covered in the study, sticking to the assumptions that all the bus lines should be studied at least once, however, in central zones at least once in all the time periods. What is more, not less than 250 km should have been covered during one time period. The measurements were carried out personally by the author of this paper in 2017 with the use of measurement form supported by GPS logger.

GPS tracks imposed on the map of moderate speed zone in central area of the city of Jaworzno (speed limit up to 50 km/h) are presented in Fig. 2.

In moderate speed zones in central area of the city (speed limit up to 50 km/h), according to the measurements, the values were as follows:

$$w_{Z_1} = \begin{cases} w_{Z_1}^{tp_1} = 1.29 \text{ in the time period } 05^{00}-09^{00} \\ w_{Z_1}^{tp_2} = 1.95 \text{ in the time period } 09^{00}-13^{00} \\ w_{Z_1}^{tp_3} = 2.87 \text{ in the time period } 13^{00}-17^{00} \\ w_{Z_1}^{tp_4} = 1.40 \text{ in the time period } 17^{00}-21^{00} \end{cases} \tag{3}$$

In moderate speed zones in suburban area of the city (speed limit up to 50 km/h), the values were as follows:

$$w_{Z_2} = \begin{cases} w_{Z_2}^{tp_1} = 0.45 \text{ in the time period } 05^{00}-09^{00} \\ w_{Z_2}^{tp_2} = 0.47 \text{ in the time period } 09^{00}-13^{00} \\ w_{Z_2}^{tp_3} = 0.56 \text{ in the time period } 13^{00\cdot}-17^{00} \\ w_{Z_2}^{tp_4} = 0.47 \text{ in the time period } 17^{00}-21^{00} \end{cases} \tag{4}$$

In the areas of increased speed (speed limit equal and above 60 km/h), the values were as follows:

$$w_{Z_3} = \begin{cases} w_{Z_3}^{tp_1} = 0.19 \text{ in the time period } 05^{00}-09^{00} \\ w_{Z_3}^{tp_2} = 0.02 \text{ in the time period } 09^{00}-13^{00} \\ w_{Z_3}^{tp_3} = 0.21 \text{ in the time period } 13^{00\cdot}-17^{00} \\ w_{Z_3}^{tp_4} = 0.21 \text{ in the time period } 17^{00}-21^{00} \end{cases} \tag{5}$$

Figure 3 is a graphical illustration of the test results in the form of a bar chart. The greatest values of the indicator of stops of buses in the traffic speed zones $w_{z_j}$ in relation to the number of kilometers driven is observed in moderate speed zones in central area of the city. These values (1.29 ÷ 2.87) are significantly higher than in the others traffic speed zones (0.45 ÷ 0.56 and 0.02 ÷ 0.21). This state of affairs is caused due to much worse traffic conditions in the city center and the resulting large number of stops of the bus, which do not result from the timetable. The main reasons for stopping were pedestrian crossings, 'give way' and 'stop' traffic signs and traffic jams.

**Fig. 2.** GPS tracks imposed on the moderate speed zone in central area of the city of Jaworzno (Source: own based on [18])

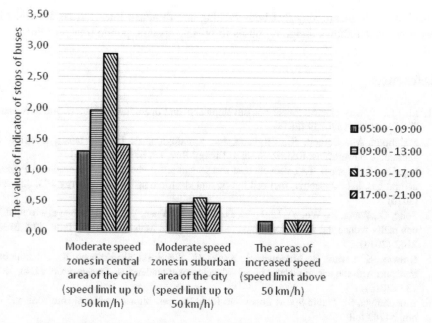

**Fig. 3.** The values of indicator of stops of buses in the traffic speed zones

In terms of time, the biggest fluctuations of the indicator of stops of buses in the traffic speed zones $w_{z_j}$ in relation to the number of kilometers driven take place in moderate speed zones in central area of the city with speed limit up to 50 km/h. In the other two traffic speed zones variations over time are negligible. The largest number of bus stops were noticed during the afternoon peak. The relatively low value of the number of bus stops during the morning rush can be a surprise, as in Polish conditions traffic intensity in the morning and afternoon rush hours is quite similar. Such measurement results may result from significantly higher pedestrian traffic resulting from the opening hours of numerous points of interest (shops, offices, etc.).

## 5  Summary

The indicator of stops of buses in the traffic speed zones $w_{z_j}$ in relation to the number of kilometers driven can be an interesting alternative describing the route of a bus line. This indicator allows assessing the ecological aspects of the implementation of electric buses. The higher is the number of theoretical stops per line route, the more harmful emissions are avoided by using electric buses. When looking for the most susceptible lines to be operated by electric buses, one should search for the line with the highest value of the theoretical number of stops over the line.

The presented indicator of stops of buses in the traffic speed zones $w_{z_j}$ in relation to the number of kilometers driven, is important because it allows rational, gradual introduction of electric buses, starting from those lines that are the most susceptible for this. It is worth mentioning that both existing and designed lines can be described in this way, which allows assigning buses to operate specific public transport lines.

# References

1. Li, J.Q.: Battery-electric transit bus developments and operations: a review. Int. J. Sustain. Transp. **10**(3), 157–159 (2014)
2. Mahmoud, M., Garnett, R., Ferguson, M., Kanaroglou, P.: Electric buses: a review of alternative powertrains. Renew. Sustain. Energy Rev. **62**, 673–684 (2016)
3. Lajunen, A., Lipman, T.: Lifecycle cost assessment and carbon dioxide emissions of diesel, natural gas, hybrid electric, fuel cell hybrid and electric transit buses. Energy **106**, 329–342 (2016)
4. Song, Q., Wang, Z., Wu, Y., Li, J., Yi, D., Duan, H., Yuan, W.: Could urban electric public bus really reduce the GHG emissions? A case study in Macau. J. Clean. Prod. **172**, 2133–2142 (2018)
5. Krawiec, S., Łazarz, B., Markusik, S., Karoń, G., Sierpiński, G., Krawiec, K.: Urban public transport with the use of electric buses - development tendencies. Transp. Prob. **11**(4), 127–137 (2016)
6. Ameratunga, S.: Traffic Speed Zones and Road Injuries. https://www.bmj.com/content/339/bmj.b4743.full
7. Chen, P.: Built environment factors in explaining the automobile-involved bicycle crash frequencies: a spatial statistic approach. Saf. Sci. **79**, 336–343 (2015)
8. Steinbach, R., Grundy, C., Edwards, P., Wilkinson, P., Green, J.: The impact of 20 Mph traffic speed zones on inequalities in road casualties in London. J. Epidemiol. Commun. Health **65**(10), 921–926 (2011)
9. Macioszek, E., Czerniakowski, M.: Road traffic safety-related changes introduced on T. Kościuszki and Królowej Jadwigi streets in Dąbrowa Górnicza between 2006 and 2015. Sci. J. Silesian Univ. Technol. Ser. Transp. **96**, 95–104 (2017)
10. Macioszek, E., Lach, D.: Analysis of the results of general traffic measurements in the West Pomeranian Voivodeship from 2005 to 2015. Sci. J. Silesian Univ. Technol. Ser. Transp. **97**, 93–104 (2017)
11. Macioszek, E., Lach, D.: Analysis of traffic conditions at the Brzezinska and Nowochrzanowska intersection in Myslowice (Silesian Province, Poland). Sci. J. Silesian Univ. Technol. Ser. Transp. **98**, 81–88 (2018)
12. Szczuraszek, T.: Bezpieczeństwo Ruchu Miejskiego. WKiŁ, Warsaw (2005)
13. Shrestha, K.J., Shrestha, P.P.: Comprehensive framework for speed-zone guidelines. J. Traffic Transp. Eng. **3**(4), 352–363 (2016)
14. Van Mierlo, J.V., Maggetto, G., Van de Burgwal, G., Gense, R.: Driving style and traffic measures-influence on vehicle emissions and fuel consumption. Proc. Inst. Mech. Eng., Part D: J. Automobile Eng. **218**(1), 43–50 (2004)
15. Ortuzar, J., Willumsen, L.: Modelling Transport. Wiley, Chichester (2011)
16. Krawiec, K., Kłos, M.J.: Parameters of bus lines influencing the allocation of electric buses to the transport tasks. In: Macioszek, E., Sierpiński, G. (eds.) Recent Advances in Traffic Engineering for Transport Networks and Systems. LNNS, vol. 21, pp. 129–138. Springer, Switzerland (2018)
17. Rudnicki, A.: Jakość Komunikacji Miejskiej. SITK, Krakow (1999)
18. QGIS. https://www.qgis.org/pl/site/forusers/

# Modelling Tools in Service
# of Development of Traffic Engineering

# Assumptions to Application of Address Points as Traffic Zones in Modelling Travel in Cities

Jacek Chmielewski[(⊠)] and Tomasz Szczuraszek

Faculty of Construction, Architecture and Environmental Engineering,
University of Technology and Life Sciences, Bydgoszcz, Poland
{jacek-ch, zikwb}@utp.edu.pl

**Abstract.** The paper presents the assumptions and construction of an innovative travel demand model, developed for the needs of traffic analysis in a small town in Poland, including prognostic analyses. In the model the classical division of the analysis area into transportation zones was abandoned, and instead the city's address point database was used to define the origins and destinations not a transportation zones, but as specific buildings located in the analysed area. In addition to the description of exchange trips, the superior travel demand models of areas covering districts, provinces and the whole country were used. In the model, all forms of transport have been reconstituted: walking, cycling and public transport by train and bus. Verification of its correctness by GEH statistic comparing the traffic volume from traffic measurements against that obtained from the transport model proofed its correctness. In addition, the model was verified by statistical hypotheses about the compatibility of trip length distributions obtained from surveys of inhabitants and those from the model.

**Keywords:** Transport model · Travel demand · Transportation zones

## 1 Introduction

Transportation as an everyday phenomenon, comprising the movement of both people and goods, is a complex process, dependent on a number of factors. It is not easy, therefore, to describe it in a standardized manner. Some factors, often of a random nature, affect the decision-making process of transportation network users as regards the realization of a trip or journey from the travel origin to the destination using a selected means of transport and one of numerous routes, and thus essentially make it impossible to accurately reproduce all the occurrences within transportation systems. Nevertheless, a more and more reliable description of the transportation reality is pursued through more precise mathematical interpretation of individual occurrences affecting everyday movement of passengers and goods. However, even for a small transportation network, such an approach requires the application of appropriate IT tools to support the analysis of transportation processes.

When reproducing traffic phenomena it is particularly essential to correctly define both the transport demand (the residents' transport demands) and the supply (the availability of transportation systems). It is noteworthy that transport demand is not merely a consequence of natural living needs of residents; it concerns journeys

© Springer Nature Switzerland AG 2019
E. Macioszek and G. Sierpiński (Eds.): Directions of Development of Transport
Networks and Traffic Engineering, LNNS 51, pp. 153–162, 2019.
https://doi.org/10.1007/978-3-319-98615-9_14

undertaken both by the residents of an analyzed area (e.g. intra-city trips and travelling beyond the administrative boundaries – generated trips) and by people living outside the area. The latter includes trips made by visitors running errands in the city and transit journeys the origin and the destination of which are located elsewhere but which use the transport infrastructure of the analyzed area.

A correct description of internal and external (exchange) trips in travel demand models makes it possible to be used in the analysis of the functioning of individual transportation systems, which is the main purpose of development of such models. Undoubtedly, one of the key obstacles to achieve a high degree of precision of these models is the accuracy of the description of the so-called travel demand layer.

This paper presents a description of the assumptions and structure of an innovative travel demand model, which uses address points as the sources of generation and absorption of most trips within the city. The object of research, for which such a model was developed for the first time, is a small town of Nakło nad Notecią with a population of about 19 000, a population density of about 1 800 people per square kilometer, and a total area of 10.65 km². The town is situated in Mid-West Kujawsko-Pomorskie province, about 25 km in a straight line from the provincial capital and a large economic centre, Bydgoszcz.

The other innovative feature of the travel demand model presented in this paper is its hierarchical relationship with higher-level models, i.e. regional, provincial and national. This model applies to the local level; it is the most detailed of the four, however a lot of data used for the development, especially generated, absorbed and transit traffic data, was sourced from the regional model - the so-called Bydgoszcz-Toruń Functional Area (B-TOF), and from the provincial and national models.

The practical objective of the development of this travel demand model was to identify potential possibilities of improvement of transportation systems by introducing changes to the organization of traffic in the town and to set out future possibilities of the town's spatial development. The scientific objective of the paper is to implement a novel method of definition of travel origins and destinations in the modelling of transportation processes in order to enable a much more detailed representation of these processes in comparison to conventional transportation zones [1], and to implement superior travel demand models to define the so-called external transportation zones which will provide a more precise description of exchange trips in the analyzed town.

## 2    Description of the Research Methods

The elaboration of a reliable travel demand model required appropriate and detailed source data describing the spatial development of the analyzed area, its transport infrastructure, the performance of the transportation systems, traffic volumes and passenger flows, the capability of individual facilities (address points) to generate and absorb trips, the residents' travel behaviors and the like. For the development of the model the required data and information are obtained from various sources, including through direct field studies, which included the following:

- observations of external traffic (exchange trips),
- observations of internal traffic (internal trips),
- surveys of the residents.

Typically road traffic counts should be carried out from Tuesday to Thursday, because the mid-week days demonstrate the most average travel behaviors of the residents.

The observations of external road traffic consisted in a cordon count should be used to determine the matrix of transit traffic through the town, as well as the incoming and the outgoing traffic. It should involve making records of all vehicles moving into and out of the town at all entry and exit roads; the type of every vehicle, the last 5 characters of its registration mark and the time of entry/exit should be recorded. The observation of exchange trips by public transport should involve counting the actual passenger numbers on all public transport means of transport like buses, trains etc. The data are necessary to identify passenger flows in individual public transport lines and movement corridors which provide the base material for the calibration of a travel demand model. The occupancy of the means of public transport may be checked using different methods, depending on the means of transport. For example in the case of railway transport, the counts may be performed on trains during journeys between individual stations. In the case of buses, observations may be carried out from the outside.

The observations of internal road traffic consisted in the measurement of traffic flows on a number of selected urban roads and included the identification of different vehicle classes. The observations may involve the following two methods:

- manual counting by specially trained staff, carried out from 6 to 10 am on selected days of the week,
- automatic counting by for example Viacount II traffic counters that can be used for automatic measurement of vehicular traffic flows including vehicle length and speed; these count may be carried out for few days non-stop; where on the base of the length data the vehicles maybe classified into individual classes.

Another crucial method is a survey among the residents. The main purpose of the survey is to identify key data for the determination of applicable travel demand model parameters. An indirect objective, although leading to the determination of the parameters, is to identify a set of characteristics describing the patterns and frequency of trips done by different groups of transportation network users, as well as their choice of destination and means of transport. The elementary information on the residents' travel behavior is obtained from interviews carried out in their households. For the purpose of the survey, questionnaire forms should be developed which consist of two parts: a general profile of the respondent and a so-called 'travel log' (trips diary). The questionnaire should be is designed to allow identification of all the characteristics of the respondents' travel behavior which are required to develop a travel demand model. The questionnaire should ask the respondents to describe their typical weekday (Tuesday, Wednesday or Thursday) directly preceding the day of completion of the form. Trained interviewers should visit preselected households where they interviewed each family member aged 9 or more (it is assumed children under 9 do not travel alone). The correctness of the responses ought to be verified using different check tools, like a database and dedicated software.

## 3   Assumptions for the Development of the Travel Demand Model

The presented travel demand model, based on an address database for the analysed area, was developed according to the following key assumptions as regards its functions:

(1) The travel demand model is a local one, spatially referenced to the National Coordinate System 2000, Zone 6, being a hierarchical subdivision from superior travel demand models, i.e. from the regional Partnership Area model, which in turn is derived from a regional travel demand model covering 1/16 of the National Travel Demand Model. The analysed model provides the greatest amount of detail as compared to the superior models, which it is based on for the purpose of identification of exchange and crossing trips and those absorbed and generated by the town. Attention is drawn here to the fact that such a hierarchical division of travel demand models enables consideration of supralocal occurrences, which affect local transportation systems, in a local model. For example, the construction of an expressway will provide a significant relief for the traffic on local roads.

(2) The travel demand model is designed on the basis of two sub-models, and each of them consists of two layers: supply and demand, and calculation procedures which describe the interactions between them. The first sub-model describes internal trips undertaken within the boundaries of the analysed area, whereas the other one deals with external (exchange) trips, which include trough trips, and absorbed and generated ones.

(3) The supply layer of internal trips is described using a classic directed graph. The inclusion of all transportation networks in the model: road sections of all levels of service, sidewalks (mainly bus stop and station access pathways, park crossings and front door paths), bikeways, and public transport systems: railways and bus services, is an innovative approach. It results from the local character of the travel demand model and its exceptional degree of detail. The data on the geometry of individual sections of the transport infrastructure presented in the graph maybe sourced from a public domain database - OpenStreets - in the Shapefile format (shp) [2], and then validated on the basis of field visits and using Google Street View (basic properties of individual sections and intersections: numbers of traffic lanes and their width, speed limits, kind and condition of the road surface, type of intersection, road hierarchy etc.). As there is no internal public transport within the analysed area, the model comprised all railway and bus lines serving the public in the area (routes, travel times, stops and the number of runs in 24 h).

(4) The demand layer is defined in an non-standard way. As already mentioned, considering the assumed high precision of the analysed travel demand model, the conventional identification of transportation zone boundaries, defined as areas representing homogenous land use [1], was rejected. Instead, according to actual

origins and destinations of trips undertaken by residents and visitors, buildings located within the administrative boundaries of the town are assumed for this layer; they comprised residential buildings, workplaces, places of study and various forms of urban activity, including commerce, services, culture, rest and leisure, administration, education, sport and so on (see: Fig. 1). At the same time, only those buildings should be considered, for which an address is available, understood as a zip code, street name and number. Thus, the maximum possible number of origins and destinations is obtained, allowing the model to include a more dense road network [1]. For example in described travel demand model 1.973 such buildings are defined altogether. For each of the buildings no more than one connector maybe defined to the public and private transportation network. It is assumed that the tenants of the residential buildings has one parking space each and that the buildings has one entrance each. This approach best represents the actual way in which the residents undertake trips and significantly differs from conventional solutions, in which centroids of transportation zones are linked to a number of transportation network nodes, indicating potential initial and terminal trip points [3].

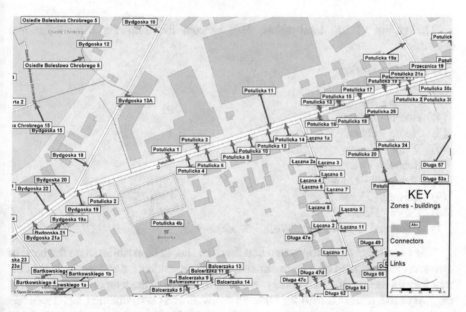

**Fig. 1.** Definition of origins and destinations of trips in the travel demand model and the connectors to the transportation network

(5) The correlation between the demand layer and the supply layer within the category of internal trips, and their mutual interactions, are modelled using a four step calculation approach [4], in which transportation phenomena are described by the following components:

- travel demand (trip generation):

Number of trips generated by residents of individual buildings $i$ motivated by a given reason $m$ in a given time interval $t$ (e.g. a day or morning rush hour):

$$P = KR \cdot \sum_{i=1}^{R} \sum_{j=1}^{G} \sum_{k=1}^{MP} R_{jk} M_{ij} \tag{1}$$

where:
$P$    - total number of internal trips,
$KR$   - mobility adjustment rate,
$R_{jk}$   - number of trip generated by an average resident falling within the category of people with homogeneous travel behaviours $j$ motivated by $k$,
$M_{ij}$   - number of residents falling within the category of people demonstrating a homogeneous transportation behaviour $j$ living in transportation zone $i$,
$R$    - number of internal transportation zones within the analysed area,
$G$    - number of groups of people with homogeneous travel behaviours,
$MP$  - number of reasons motivating the residents.

The groups of trip motivations, $MP$, concern the following trip origins and destinations:

(1)  Home - Work,
(2)  Home - School,
(3)  Home - Shopping,
(4)  Home - Leisure,
(5)  Home - Other,
(6)  Work - Home,
(7)  School - Home,
(8)  Shopping - Home,
(9)  Leisure - Home,
(10)  Other - Home,
(11)  Unrelated to home (e.g. Work - Work).

The parameters of the trip generation function should be established on the basis of own surveys conducted on a representative sample of residents, that are asked about the way they do their trips on a workday. Keeping in mind that the temporal distribution of the size of the generated trips for different motivations is both different and dependent on the group of people with homogeneous travel behaviours, it is necessary to determine daily distributions and numbers of trips generated by the representatives of each groups of residents for given motivations. The data should be based on the above-mentioned surveys. The number of all trips generated by the residents of an area for a given time interval is equal to the number of residents in the group of homogeneous travel behaviours multiplied by the mobility within a given motivation for the time interval considered.

In this model the following groups of people demonstrating similar travel behaviours are proposed to be identified:

- primary schoolchildren over 9 and junior secondary school (middle-school) students,
- secondary school students,
- college and university students,
- employed,
- unemployed and all others not listed above.

The volume of generated commercial traffic (vans and trucks) maybe established on the basis of the number of people employed and the use of individual facilities located within the analysed area, and using observations of the average number of generated trips of commercial vehicles in 24 h observation of representative companies and other entities:

- trip distribution (spatial distribution of the trip, identification of the location of the building where the trip terminates). The process is realised on the basis of an elementary gravity model function determined for the residents' trips,
- mode choice - the choice of the mode of transport to be used by the residents to trip between the origin and the destination. The residents are assumed to have the following travel modes and options: by car as a driver, by car as a passenger, by public transport, on foot, by bicycle or using a combination of these (multimodal transport) i.e. by car as a driver or passenger to a P&R area and then by public transport. In the case of commercial vehicles the mode choice is made at the time of trip generation,
- traffic assignment into the transportation network (the choice of route using a particular transportation mode and the choice of subsequent streets when walking from the origin to the destination).

Vehicle traffic, whether it concerns cars, commercial vans or trucks analyses, are performed on the basis of the distribution of sustainable user cost - the User Equilibrium Assignment [5], based on Wardrop's first principle which states that no driver may unilaterally reduce the cost of travel by choosing an alternate route.

(6) The supply layer in the exchange trip sub-model comprises roads designed for vehicle traffic and railway lines crossing the boundary of the analysed area. These objects are strictly correlated with the supply layer of internal trips (see: 3).

(7) The demand layer in the exchange trip sub-model comprises trips undertaken by car and using external public transport whose routes are included in the analysed model area. Such trips are generated by the residents of the analysed area (trips related to work, study etc.), absorbed by the analysed area (visits to the area for a specific motivation) and trips through the urban area. For the purpose of a conventional definition of this layer, at individual entries to/exits from the modelled area so-called external transportation zones [4] should be defined, representing transport corridors (roads, railway and bus lines) allowing people to travel to/out of and through the analysed area.

(8) Similarly to the first sub-model, the link of the demand layer with the supply layer for exchange trips and their mutual interactions are modelled on the basis of a three-step calculation model:

- trip generation - data on the number of through trips, absorbed and generated trips, divided into trips made by car (as the driver or a car passenger) and by public transport, as well as light and heavy commercial trips at each entry to and exit from the analysed area should be obtained using the superior level travel demand models: the regional and the national one. The number of through trips typically is assumed to be constant regardless of the traffic appeal of the town. For external transportation zones variables should be defined to describe the number of people leaving the analysed area (generated trips) and arriving (absorbed trips). It should be assumed that each person assigned to one of the groups generates two trips on a day travelling out of the area and back or arriving in the area and leaving it. This means that each person coming into the analysed area leaves it within 24 h (the period of model operation),
- in the case of trips generate by the area one type of motivation is assumed: other trips, whereas in the case of trips absorbed by the area the considered motivations are: work, study, leisure and other trips. The generation of transport demands related to commercial traffic, i.e. light and heavy trucks traffic, is effected in a comparable manner to the internal traffic,
- trip distribution (spatial distribution of the trip, identification of the location of the building where the trip terminates) in the case of absorbed and generated trips is established similarly to the first sub-model, on the basis of the appeal of individual internal transportation zones (for absorbed trips) and external transportation zones (for generated trips), and using the matrix of mean times and distances between them,
- the distribution determines the spatial location of destinations. For the spatial distribution of the traffic a gravity model is used with a constrained role of the trip time and distance attributes,
- traffic assignment - the last of the calculation phases, realised simultaneously with the analogous phase for the internal traffic sub-model.

## 4  Model Calculation Steps

Simulation calculations according to above mentioned assumption can be performed using Procedures and Functions of the PTV VISUM software [6]. The procedures related to trip generation in both sub-models are realized on a one-off basis, according to the principles defined above. The other calculations, related both to the spatial distribution of a trip, the mode choice and the traffic assignment, are functions dependent on the transportation networks saturation (traffic volume divided by transport system capacity), and thus the functions of utility of available transportation systems. Therefore, iterative calculations are performed as part of the analysis in which - as the first iterative step - the demand part is generated with an assumed lack of traffic on the transportation networks and its full availability.

In the subsequent steps, the utility of individual transportation systems is updated. As mentioned earlier, it is understood as a general cost of use of a particular transportation mode to realize a trip [7]. It is defined by equivalent travel time by means of the particular mode - for example: the time required to access the vehicle, trip duration on a particular route, cost of travel/fare and parking, and the time required to leave the vehicle (egress) and reach the destination.

The following elements are considered: the loading of the transportation network with the ensuing loss of time being a result of the degree of saturation of the transportation systems (traffic loads on road sections and intersections, occupancy of the means of collective public transport), the resulting discomfort, extension of travel time or limited availability. At every iterative step the utility of individual transportation systems is updated and new calculations are made to determine the spatial trip distribution, transportation mode choice and transportation demand assignment. The calculations are complete when the difference of the state of network volume in two subsequent iterative steps is smaller than the assumed difference (for the purpose of the modelling the difference is assumed not to exceed 1%) or when the maximum number of iterations assumed in an expert way is reached (in the described analysis a maximum of 15 iterative steps are assumed). As a noteworthy fact, the described structure of the model and the applied system of calculations makes it possible to supplement the model with any other transportation system, having defined its utility function and considered necessary parameters of the mode choice and traffic assignment parameters. Furthermore, the utility function of a given transportation mode is a component of a number of parameters affecting its appeal to travellers (travel time, direct costs, extra fees such as parking fees or city centre access fee, the availability of transport services etc.). Thus, it provides the possibility of considering different attributes of individual transportation systems and the effects of implementation of measures aiming at a reduction of vehicle traffic, e.g. paid parking zones.

The ultimate effects of the described travel demand model include, for example, maps of utilization of transportation systems: the volumes of vehicle, bicycle and pedestrian traffic and passenger flows of external public transports at individual sections of the transportation network, as well as a number of transportation characteristics required for economic studies of the operation of transportation systems (e.g. performance, average speed, time spent travelling).

The above-described calculation procedure is rather time-consuming. In the case of the presented travel demand model the calculations did not take longer than 30 min, however the model applies to a small town. Therefore, considerably more time is expected to make all the calculations for a larger area.

## 5 Conclusions

The presented assumption for a travel demand model construction in which the origins and destinations of trips are represented by address points, rather than by conventional transportation zones, developed for a small town, features a considerable amount of detail and accuracy. A major advantage of the model is its very accurate database concerning the land use of the analysed area and a high precision of reproduction of the residents' trips. The exact location of the trip origins and destinations and their connection with the transportation networks using single 'connectors' provides clarity, transparency and accuracy of the representation of traffic phenomena occurring in real-world conditions. This applies to car trips, but also to pedestrian and bicycle traffic, as well as to trips undertaken by means of public transports. At the same time, it is noteworthy that the share of pedestrian journeys within cities ranges from about 20% in large cities (18% in Warsaw [8], as a consequence of a considerable distances between origins and destinations) to approximately 50% in small towns (49.98% in the analysed case). Such journeys are usually up to 0.9 km and they are normally underestimated in conventional travel demand models. By analogy, pedestrian journeys made to access public transport facilities in conventional modelling are also largely averaged (it is usually time as a function of distance from a network node, established for the whole transportation zone). In the presented model, on the other hand, these journeys are represented by actual distances between an address point (building) and a stop or station of a public transport service.

Considering the on-going increase of computing power and data storage capacity, the possibility of storage and processing of large amounts of data should not be an obstacle for the kind of solution presented in this paper. Thus, such travel demand models have a good chance of further development.

# References

1. Cascetta, E.: Transportation Systems Analysis. Models and Applications. Springer Optimization and Its Applications, vol. 29. Springer US, New York (2009)
2. OpenStreets Maps. https://www.geofabrik.de/data/download.htmh
3. Chang, K., Khatib, Z., Ou, Y.: Effects of zoning structure and network detail on traffic demand modeling. Environ. Plan. B: Plan. Des. **29**, 37–52 (2002)
4. McNally, M.: The Four Step Model. https://www.its.uci.edu/its/publications/papers/CASA/UCI-ITS-AS-WP-07-2.pdf
5. Bellei, G., Gentile, G., Papola, N.: A within-day dynamic traffic assignment model for urban road networks. Transp. Res. Part B **39**, 1–29 (2005)
6. PTV VISUM Manual. http://vision-traffic.ptvgroup.com/faq-files/Installation_Visum16.pdf
7. Mathew, T., Rao, K.: Introduction to Transportation Engineering. https://www.civil.iitb.ac.in/tvm/2802-latex/demo/tptnEngg.pdf
8. Kostelecka, A.: Report from the 3rd Stage of the Warsaw Transportation Research. Technical Report. Capital City of Warsaw, Warsaw (2015)

# Monitoring Urban Traffic from Floating Car Data (FCD): Using Speed or a Los-Based State Measure

Oruc Altintasi$^{(\boxtimes)}$, Hediye Tuydes-Yaman, and Kagan Tuncay

Faculty of Civil Engineering, Middle East Technical University, Ankara, Turkey
{aoruc,htuydes,tuncay}@metu.edu.tr

**Abstract.** Floating Car Data (FCD) has an important traffic data source due to its lower cost and higher coverage despite its reliability problems. FCD obtained from GPS equipped vehicles can provide speed data for many segments in real-time, as provided by Be-Mobile for urban regions in Turkey. Though only providing speed per consecutive road segments, FCD is a great data source to visualize urban traffic state, in the absence of any other traffic data source, which is the focus of this study. After evaluation of variations of FCD speed values, a more simplified but more robust measure, called traffic state level (TSL) was proposed based on the Level of Service definition for urban arterials in the Highway Capacity Manual. Numerical results from analysis of one month FCD from May 2016, showed the capability of FCD and advantages and limitations of visualization based on TSL measure, which can be very efficient in data archiving, as well.

**Keywords:** Traffic pattern detection · Traffic state estimation
Data visualization · Floating Car Data

## 1 Introduction

Accurate and reliable estimation of the traffic state in urban arterials is a crucial step of traffic management and control. Traditionally, various traffic data sources (magnetic loops, road tube counters, radars, Bluetooth) have been used to estimate different traffic parameters (i.e. link occupancy, average speed or density of a corridor), which are eventually combined to estimate traffic state (or characteristics) based on the fundamental relationship between flow-density-speed. When this relation is obtained, it is possible to identify different traffic states (congested, free flow, etc.) directly.

More recently, Floating Car Data (FCD), has been used increasingly in traffic studies, mainly due to its lower cost and higher coverage, despite its reliability problems. The principle of FCD is to collect real-time traffic data by locating a vehicle via mobile phones or GPS over the entire road network. Data such as car location, speed and direction of travel are sent anonymously to a central processing center, which is later processed to derive travel time or average speeds of road segments. If it is from a single probe vehicle, speed and position data of the floating cars can be obtained periodically (e.g.1 min or 5 min) for each road segment [1]. However, data from a

© Springer Nature Switzerland AG 2019
E. Macioszek and G. Sierpiński (Eds.): Directions of Development of Transport
Networks and Traffic Engineering, LNNS 51, pp. 163–173, 2019.
https://doi.org/10.1007/978-3-319-98615-9_15

vehicle fleet is generally preprocessed to obtain segment speed (or travel time) only. Even though GPS usage and affordability has increased, it has been mostly used to monitor typically fleet management services (such as, taxi drivers, trucks). While traffic data obtained from private vehicles or trucks are more suitable for motorways and rural areas, taxi fleets are particularly useful due to their high number and their on-board communication systems in urban regions [2].

FCD for Turkey comes from 600 000 GPS equipped vehicles (among the total 19 million registered vehicles), which corresponds to approximately 3% penetration rate. With such penetration, FCD has become a major traffic data source for urban arterials, along which there existed almost no traffic data sources or counts before. Today, real-time broadcasted FCD has provided traffic data for almost 5 000 segments (road segments less than 50 m) in Ankara, which covers almost 250 km of urban road network, and requires storage capacity of 8 GB area per day. As it is not possible or meaningful to store the FCD data every day, it has to be analyzed to derive and visualize traffic states in a realistic and practical way, which is the main focus of this study.

The ultimate objective is defined as assessment of the power of FCD to visualize the time-dependent nature of traffic flow characteristics, and which parameters to employ in the process. Two alternatives tried included the speed values published real-time in FCD, which is more precise, and a more simplified state based measure modified from Level of Service (LOS) concept proposed for urban corridors in the Highway Capacity Manual [3]. Numerical results were developed based on the evaluation of FCD on a major urban arterial archived for the May, 2016. Any discussion on the validity or verification of FCD is avoided in the absence of density or flow values along the corridor, yet.

## 2  Literature Review

### 2.1  Traffic Data Visualization

Visualization of the traffic patterns has been performed either using historical data set or using real time traffic data [4]. Historical data analysis covers the examination of the speed distribution, statistical methods such as clustering, principal component analysis, etc. to capture the traffic patterns. Pongnumkul et al. [5] used historical data set for examining the speed profiles and defined different congestion levels based on the different speed values to visualize the road network. Vasudevan et al. [6] defined the congestion state based on the decrease in speed in accordance with the free flow speed. While, highly congested state was defined as the speed less than the 1/3 of the free flow speed, uncongested state was specified as the speed greater than 2/3 of the free flow speed. Pascale et al. [7] proposed the spatio-temporal clustering methods for the visualization of the traffic patterns in urban arterials. The well-known fundamental diagram showing the relation between the flow-speed-density was modified for the urban arterials. Clustering analysis resulted in 6 different states and threshold speed values were determined for each state. Later, selected urban arterial was visualized according the determined speed thresholds. On the other hand, Liu et al. [8] defined speed threshold values for 5 different traffic states, but they did not discuss how they obtained these threshold values. Li et al. [9] used 3 months of historical FCD to examine variabilities in average speeds, and attempted to determine congestion locations depending on sudden decreases in average speeds in consecutive road segments. Similar to this study, Xu et al. [1] highlighted the issues when dealing with the enormous historical data set when endeavoring to find meaningful traffic and congestion patterns. They obtained FCD from 12 000 GPS-equipped taxi fleets in Wuhan city, China. They proposed a statistical method for data analysis (data cube management) for congestion pattern detection and visualization. In contrast to these studies, Adu-Gyamfi and Sharma [10] explored the reliability of probe speed data for detecting congestion trends. The study focused on pattern recognition and time series data analysis to identify similarities with probe-based speed data. Altintasi et al. [11] also proposed pattern search algorithm to detect 12 different traffic patterns (i.e. bottleneck release locations, congested flow, stable flow conditions etc.) from FCD.

## 2.2   LOS for Urban Arterials

LOS is a quantitative measure representing quality of service [3]. Generally, 6 different LOS states are defined for different road types, where LOS A represents the best operating condition and LOS F the worst. While the main parameter for the LOS for freeways and motorways is density, this parameter cannot be a direct measure of the urban arterial LOS. HCM [3] defined LOS for urban arterials as "the reductions in travel speed as a percentage of the free-flow speed of the corridor". Table 1 shows speed reduction thresholds as stated in the HCM, and corresponding speed intervals for urban arterials with speed limits of 50 km/h and 90 km/h separately.

In Turkey, urban arterial roads have a legal speed limit of 50 km/hr, which can be increased up to 70 km/h or even 82 km/h (which corresponds to 90 km/h in practice by a 10% tolerance margin) by local government. As the posted speed limit on the study corridor is 82 km/h, LOS definitions for 90 km/h free-flow are assumed where LOS A and LOS B corresponded to speed intervals of "90 km/h–77 km/h" and "77 km/h–60 km/h", respectively. Speed intervals for flowing conditions represented by LOS C and LOS D were assumed as to "45 km/h–60 km/h" and "36 km/h–45 km/h", respectively. Finally, a segment speed less than 27 km/h was the worst case, resulting in LOS F.

**Table 1.** Average speed and corresponding LOS values for the urban roads based on [3].

| LOS | Travel speed as a percentage of base free-flow speed | Travel speed intervals for speed limits of | | Proposed traffic state level (TSL) |
|-----|-----|-----|-----|-----|
| | | 50 km/h | 90 km/h | |
| A | >85 | >43 | >76 | 1 |
| B | 67–85 | 34 < V< 43 | 60 < V < 76 | |
| C | 50–67 | 25 < V < 34 | 45 < V < 60 | 2 |
| D | 40–50 | 20  < V < 25 | 36 < V < 45 | 3 |
| E | 30–40 | 15  < V < 20 | 27 < V < 36 | 4 |
| F | <30 | <15 | <27 | |

# 3 Study Area and FCD

## 3.1 FCD Structure

FCD data used in this study is provided by Be-Mobile, a Belgium-based traffic information provider, which delivers real-time average speed data at one-minute intervals for road segments of lengths shorter than 50 m. FCD includes two sets of tables (see Tables 2 and 3). While static table included the static characteristics of the road segments (such as, segment ID, length, road class type, speed limit, average speed, and the coordinates), dynamic tables included attributes varying over time (such as, day, time based average speed values, travel time, # of probe vehicle passed at a given time etc.). Segment ID was repeated in both tables as the unique identification code.

**Table 2.** Static information of sample road segments located in the study corridor.

Static FCD data structure

| Segment Id | Length [m] | Type | Avg. speed [km/h] | Speed limit [km/h] | Start coord. | End coord. | Local Id |
|---|---|---|---|---|---|---|---|
| 1215926 | 49.24 | 5 | 60 | 70 | 32.8134, 39.9184 | 32.8134, 39.9179 | 1 |
| 1215927 | 49.24 | 5 | 60 | 70 | 32.8134, 39.9179 | 32.8135, 39.9175 | 2 |
| 1215928 | 49.24 | 5 | 60 | 70 | 32.8135, 39.9175 | 32.8135, 39.9170 | 3 |
| 1215933 | 48.01 | 5 | 60 | 70 | 32.8135, 39.9170 | 32.8135, 39.9166 | 4 |
| .... | | | | | | | |

In this study, average speed values for each segment are recorder at every 1-min interval. However, Be-mobile publishes the average speed data as truncated at the road speed limits. Though the speed limit of the study corridor is currently increased to 82 km/h (and 90 km/h in practice due to the tolerance in the enforcement) by the municipality, speed data obtained from Be-mobile was truncated at previously accepted limit of 70 km/h. Though, this created some loses in the determination of probability distribution of speeds for the segments, such aggregation on the speeds close to free-flow conditions caused very little problem in the evaluation of problematic traffic conditions and states observed in the congestion limits.

**Table 3.** Dynamic information of sample road segments located in the study corridor.

Dynamic FCD data structure

| Segment Id | Day | Time [h] | Travel time [s] | Speed [km/h] | # of probed veh. |
|---|---|---|---|---|---|
| 1215928 | 01.07.2015 | 8:01 | 2.48 | 70.00 | 10 |
| 1215928 | 01.07.2015 | 8:02 | 2.48 | 70.00 | 8 |
| ... | | | | | |
| 1215928 | 01.07.2015 | 9:00 | 2.89 | 61.34 | 10 |
| ... | | | | | |

## 3.2    Study Corridor

As a case study, a 4.5 km corridor on Dumlupınar Boulevard (the stretch from the Hacettepe University interchange to Bilkent University entrance), which is a major arterial in the form of a multilane urban highway corridor (4 lanes in each direction) in Ankara, is selected as shown in Fig. 1. The study period included the data archived from May 2016 for the morning peak hour period (between 07:30–09:00) including weekdays, only. The study corridor consists of 82 segments in one direction. For these segments, local numbering has been assigned in a consecutive way instead of the Be-mobile's segment numbering (see Table 1). Study area includes one section with electronic speed enforcement point (spot speed enforcement located around the road Segment ID 27), and 2 major grade-separated interchanges, Hacettepe and Bilkent, as shown in Fig. 1.

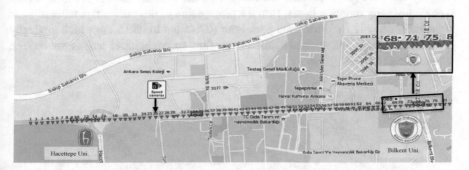

**Fig. 1.** The study corridor and the locations of the 80 road segments and close-up view of segments 68–75 located before the Bilkent interchange

# 4    Visualization of Traffic Characteristics at a Road Segment

## 4.1    Temporal Distribution of Speed in FCD in a Day

Change of FCD speed data published every minute over time was illustrated for every weekday in the first week of May, 2016 (see Fig. 2a) for Segments 68, 70 and 72. To investigate time-based evolution of traffic states, the analysis period was selected between 7:15 and 9:00 which contained the morning peak hour mostly observed between 7:45 and 8:45. Signification variations were observed on Friday, May 6th, 2016. To avoid any biased evaluation based on one week sample, weekly average speeds for all four weeks in the same month, were calculated and presented in Fig. 2b, which showed smaller variation. But, to visualize the basic descriptors of the time-dependent speed values within the month of May, 2016, graphs in Fig. 2c were prepared that displayed (a) average monthly speed at every minute, (b) upper and lower confidence intervals using 2-sigma limit, and (c) the limits of the interquartile range (Q1 and Q3 representing the 25% and 75% quartiles). As expected, confidence intervals were much larger than the interquartile range, as the mean (average speed) and the standard deviation were susceptible to any outlier in the data more than the latter. However, the interquartile range did not vary much over time, confidence intervals showed great variability, which even reached a range of ±30 km/h during the peak hour (around 8:15 am) in Segment 68. This is an expected situation during congested hours, as any small increase in the demand could push traffic condition from uncongested to congested regime and cause severe delays and slowdowns in the speeds. But, such variation is not observed nor expected during off-peak hours, such as around 7:30 am, when the demand is very low and all vehicles could travel at free flow speed almost uniformly.

Secondly, one can observe the significant change of traffic conditions at nearby locations, as across segments 68, 70 and 72 (which are 50 m long and at most 100 m away from each other). However, Segment 72 was located right before the start of the ramp of Bilkent Interchange, and faced higher densities and slowed traffic due to merging and diverging vehicles, which verified the constant slower speed observed even during early morning hours. But, the propagation of this congestion reached back to Segment 68 only for during the peak hour (07:45 am to 08:15 am which gradually improved after 8:15 and reached to almost free flow speeds at 08:45 am) as can be seen in the averages.

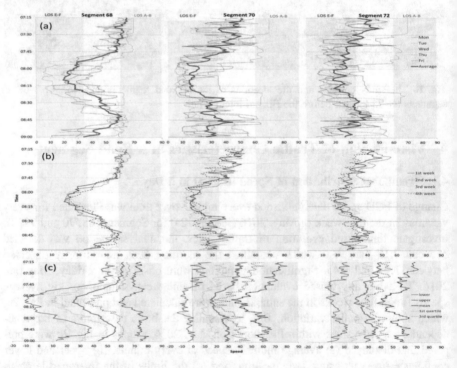

**Fig. 2.** Monitoring FCD speeds of road segments 68, 70 and 72 for (a) the first week, (b) weekly averages, (c) statistical descriptors of speed in May, 2016

## 4.2   LOS-Based Traffic State Levels in a Day

To overcome such variability visualization of traffic conditions observed when speed was used, a simpler and qualitative parameter was selected based on LOS definition in HCM [3] as discussed previously. Though LOS has 6 intervals originally, lack of speed above 70 km/hr in the existing FCD, forced an aggregation of LOS A and LOS B into one traffic state level (TSL), named TSL 1. While two levels were defined corresponding to LOS C and LOS D, as TSL 2 and TSL 3, exactly, another aggregation was assumed for the very congested regimes denoted by LOS E and LOS F, under the name of TSL 4, in this study. This TSL definition was very close to the categories employed in [6].

## 4.3   Probability Distribution of Traffic State Levels

After transforming FCD speed values to the corresponding TSL measure, time-based probability distribution of each state could be derived for the road segments as shown in Fig. 3a. On Segment 68, traffic state was at TSL 1 most of the time between 07:15–07:30, followed by TSL2 (corresponding to LOS C) during one third of the weekdays in May. The propagation of the congestion to this segment could still be observed in the increasing probability of TSL4 (corresponding to LOS E or F) between 7:45 and 8:15, with the following dissipation until 08:45 am On the other hand, the congestion started at early times and constituted the major portion of the traffic states on Segments 70 and 72, which followed the same trends observed in Fig. 2a. This showed that proposed state definitions is as capable of the capturing major traffic states without much loss despite the aggregation of the speeds.

In addition to the visualization of the 1-min TSL probability distributions, the aggregation of the TSLs for a 5-min and 15-min time period were considered for potential data archiving requirements. As an example, TSL values from 1-min FCD data were aggregated over 5-min and 15-min intervals, and probabilities of TSL values were recalculated as shown in Fig. 3b and c, respectively. The results showed that while the details of the formation of traffic states were lost to some extent when aggregated over time, the probability distribution of the TSL could still depict the start and end times of congestion formation (as seen in Segment 68) with a slight risk of over/underestimation of expected probabilities. The longer the aggregation period was the bigger the loss in the quality of visualization of traffic conditions.

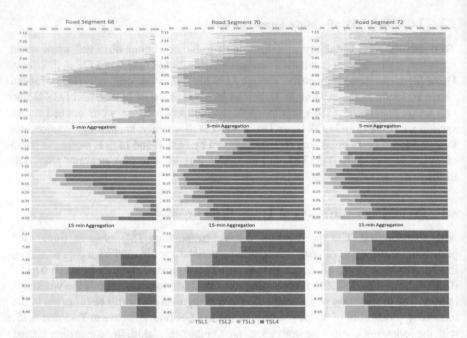

**Fig. 3.** Probability distribution of the TSLs for the road segment of 68, 70 and 72 for the time period of 07:15–08:30 (May, 2016)

# 5    Visualization of Traffic Characteristics Along Study Corridor

Visualization of TSL values calculated for each segment along a corridor consecutively would help visualization of critical traffic patterns and states, as well. However, when the horizontal axis was reserved for the segments themselves, it would not be possible to display the probability distribution of all four TSLs in addition to the time axis, nor will it be helpful to see the probabilities of each TSL. Instead, critical TSLs that were mostly observed (and thus, expected at a given time and segment) should be determined to be used in the visualization. To gain insight on the matters, first probability distribution of every TSL over time and segments was illustrated as shown in Fig. 4. While less intensely colored regions showed smaller probabilities of observance of a TSL (<25% or 25% to 50%) darker shades represented higher probabilities such as 50%–70% or >70%. For example, between Segments 1–16, traffic was predominantly at TSL1 (suggesting free flow conditions) during whole analysis period. On Segments 22 to 30, TSL 1 was predominantly observed until 07:45 am which was followed by mostly or predominantly TSL 2 until the end of the analysis period.

While TSL3 was not major along the corridor at any time, it should be remembered that it corresponded to a narrower speed interval corresponding to LOS D, which (i) is either hard to observe due to nonlinear nature of congestion formation and dissipation in traffic, or (ii) not properly defining the stable flow when LOS was made based on with single parameter as reductions from free flow speed.

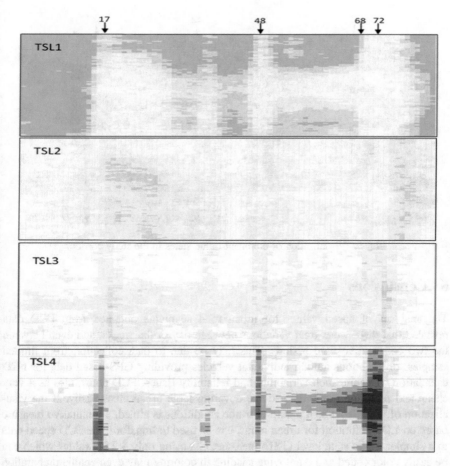

**Fig. 4.** Visualization of traffic states based on the frequency level for (a) TSL1, (b) TSL2, (c) TSL3 and (d) TSL4

Assuming the highest probability TSL values as the characteristic measure of the traffic conditions on each segment, a composite thematic map can be drawn as shown in Fig. 5. Assigning one color per TSL and two shades for each color (light one representing probability of observance of the given TSL between 50%–70% and the darker one, larger than 70%) enables a probabilistic occurrence of these conditions. Such a map showed the spatial and temporal variation of the traffic patterns more clearly. However, some segments and time intervals may not be color-coded as, they would not have a characteristics TSL value with a probability more than 50%. These are the segments and times, where traffic conditions were expected to vary much before the arrival/after the dissipation of expected congested regimes.

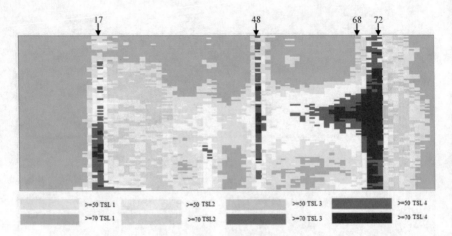

**Fig. 5.** Visualization of dominant traffic states in the study corridor

## 6 Conclusions

The analyses of speed values for urban road segments obtained from FCD data revealed that there were great variations between day of the weeks, and even between the two consecutive time periods. This may be due to data collection from limited samples, or variations among individual vehicles providing GPS based data for FCD calculations. Despite such variability and reliability issues, FCD constitutes as a very cheap and real-time traffic data source covering large urban areas easily. If the visualization of time-dependent nature of traffic conditions is aimed, a qualitative measure based on LOS definition for urban roads, can be used to transform the FCD speed data to a simpler traffic state level (TSL) measure. Choosing only 4 TSL values which can be easily color coded and employing shading in coloring may even enable preparation of probabilistic thematic maps indicating the most critical states along a corridor over a given analysis time period. Such maps can be further utilized to archive the large FCD information, where only state level is stores for the segment and time periods showing predominance of a single TSL. For time periods and segments that show great variability, even in TSL representation, probabilities of TSL values could also be stored and retrieved to reenact the realistic conditions later. However, FCD should always be integrated with any other traffic data source, providing information on other traffic parameters, density or flow, whenever and wherever possible.

Although FCD provided an opportunity for determination of traffic states in an urban corridor, its quality and reliability of such data should be evaluated which will be focused on further studies. Preliminary analysis revealed that FCD speed can be used as a surrogate measure for urban arterial speed when comparing with the ground truth speed data obtained from the specific FCD segment of the study corridor.

**Acknowledgement.** The authors would like to express their thanks to Integrated Systems & Systems Design (ISSD) company workers, for providing us to use FCD database.

# References

1. Xu, L., Yue, Y., Li, Q.: Identifying urban traffic congestion pattern from historical floating car data. Procedia – Soc. Behav. Sci. **96**(6), 2084–2095 (2013)
2. Leduc, G.: Road Traffic Data: Collection Methods and Applications. Working Papers on Energy, Transport and Climate Change. Institute for Prospective Technological Studies, Seville (2008)
3. Transportation Research Board: Highway Capacity Manual 2010. Transportation Research Board of the National Academy of Science, Washington (2010)
4. Petrovska, N., Stevanovic, A.: Traffic Congestion Analysis Visualisation Tool. https:// ieeexplore.ieee.org/document/7313335/
5. Pongnumkul, S., Kamsiriphiman, N., Poolsawas, J., Amornwat, W.: Congestion Grid: A Temporal Visualization of Road Segment Congestion Level Data. https://ieeexplore.ieee.org/ document/6645927/
6. Vasudevan, M., Negron, D., Feltz, M., Mallette, J., Wunderlich, K.: Predicting Congestion States from Basic Safety Messages Using Big Data Analytics. Transportation Research Board of the National Academies, Washington (2015)
7. Pascale, A., Mavroeidis, D., Thanh-Lam, H.: Spatio-Temporal Clustering of Urban Networks: A Real Case Scenario in London. Transportation Research Board of the National Academies, Washington (2015)
8. Liu, D., Kitamura, Y., Zeng, X., Araki, S., Kakizaki, K.: Analysis and Visualization of Traffic Conditions of Road Network by Route Bus Probe Data. https://ieeexplore.ieee.org/ document/7153888
9. Li, Q., Ge, Q., Miao, L., Qi, M.: Measuring variability of arterial road traffic condition using archived probe data. J. Transp. Syst. Eng. Inf. Technol. **12**(2), 41–46 (2012)
10. Adu-Gyamfi, Y.O., Sharma, A.: Reliability of Probe Speed Data for Detecting Congestion Trends. https://ieeexplore.ieee.org/document/7313454
11. Altintasi, O., Tuydes-Yaman, H., Tuncay, K.: Detection of urban traffic patterns from Floating Car Data (FCD). Transp. Res. Procedia **22**, 382–391 (2017)

# Problems of Deliveries in Urban Agglomeration Distribution Systems

Dariusz Pyza[1]($\boxtimes$), Ilona Jacyna-Gołda[2], and Paweł Gołda[3]

[1] Faculty of Transport, Warsaw University of Technology, Warsaw, Poland
dpz@wt.pw.edu.pl
[2] Faculty of Production Engineering, Warsaw University of Technology,
Warsaw, Poland
i.jacyna-golda@wip.pw.edu.pl
[3] Department of IT Support for Logistics, Air Force Institute of Technology,
Warsaw, Poland
pawel.golda@itwl.pl

**Abstract.** Deliveries of goods in urban agglomeration distribution systems are conditioned by many factors resulting from the specificity and character of these agglomerations. Urban agglomerations, as a specific spatial and structural system or a multi-layered spatial structure, with a specific location, spatial organization and functioning, require a systematic approach to the design of distribution systems. The article presents the characteristics of urban agglomerations and their specificity from the point of transport issues as an element of the distribution system. In addition, these factors were identified that affect the organization of deliveries and determine boundary conditions. A mathematical approach to the problem of deliveries in urban agglomeration distribution systems for the assumed decision situation was also presented. Theoretical considerations have been implemented in the case study of the urban agglomeration distribution system.

**Keywords:** Urban agglomeration · Distribution system
Mathematical model of the supply system

## 1 Introduction

The concept of agglomeration can be considered in two senses. In the first sense, agglomeration is the process of concentration or concentration in the city and its closest zone of population and economic entities. In the second sense, agglomeration is a spatially cohesive area with specific properties, resulting from the ongoing concentration process. Therefore, it can be assumed that the urban agglomeration can be both a specific spatial and structural system as well as a multi-layered spatial structure, with a specific location, spatial organization and functioning.

Two basic types of agglomerations can be distinguished: monocentric and polycentric. In the case of Poland, monocentric agglomerations are more common and typical, that is, those whose central point is one big city. There are also polycentric agglomerations in the area of the country, i.e. where the urbanized area surrounds several urban centers, a fairly diverse, often high level of centrality.

© Springer Nature Switzerland AG 2019
E. Macioszek and G. Sierpiński (Eds.): Directions of Development of Transport
Networks and Traffic Engineering, LNNS 51, pp. 174–183, 2019.
https://doi.org/10.1007/978-3-319-98615-9_16

The problem of deliveries in urban agglomeration distribution systems requires a systematic approach, taking into account both the integration of flow of goods and the associated flow of information. It is possible within the framework of an efficiently functioning logistics system of the urban agglomeration (Urban Logistics System - ULS) [1, 2]. Transport of goods is on the one hand the main factor of economic and social development of urban areas, on the other hand, it is the main source of environmental pollution and hindering the social life of inhabitants of urban agglomerations [3]. Problems related to this and solutions preventing the above phenomena are located in the area of urban logistics, whose task is to coordinate, organize and manage the flows of resources, with particular emphasis on the rationalization of flow of goods and, as a consequence, limiting freight traffic in cities.

## 2 Transport in Distribution Systems of Urban Agglomerations

Systems theory and its adaptation to the problems of urban logistics, enables the rationalization of the city's logistics system [4].

City's logistics system and its functioning, determines the occurrence of many logistic processes in it. These processes are, among others processes of supply, production and distribution of goods, as well as freight and transport processes and the flow of information. In addition, important logistics processes implemented in the city's logistics system include storage processes and ecological processes - recycling and waste disposal [5].

An important problem from the point of view of the city's functioning as a logistics system is customer service in the field of various types of consumer goods. Making the product available in a place and time that meets the needs and expectations of clients requires the design of effective distribution systems. It is especially important from the point of view of distribution systems of urban agglomerations, in which there are far-reaching restrictions resulting from the specificity and character of these agglomerations.

The distribution system creates many of its elements, and the role of individual elements in the system can vary. From this point of view, elements of the distribution system can be ordered in terms of superiority and inferiority, thus shaping hierarchical distribution systems [6–11].

In the hierarchical distribution systems, transport systems are important, the purpose of which is to optimize - in the sense of the adopted criteria - satisfying transport tasks reported in a given area [12].

The efficiency of the distribution system in cities is in a sense an indicator of the efficiency of the city itself and its management system. From the point of view of the principles of city logistics, the most important role in the distribution system is played by the transport system, which through its links affects all other subsystems of the city [5].

Considering the implementation of internal flows, the city is distinguished by the transport of people and loads. The structure of transports carried out in urban agglomerations indicates that one-third of all transport is freight transport, while two-thirds is passenger transport. Urban freight transport is identified with the delivery transport, by which is meant generally transporting freight for and in the city area, as well as for the whole agglomeration area [5].

The largest part of the transport of cargo in the city takes place using road transport, of which the largest percentage is transported by car transport. The structure and size of the carriage is determined by the type of product group, the functional areas, the amount of transported cargo, transport distance, and the types of means of transport [5, 12, 13].

The specificity of urban distribution systems causes many problems in the transport of goods in the city, such as congestion, long time passing through the city, stoppages in crowded streets and lack of parking spaces at cargo pickup points. Such a situation causes that in the transport service of urban distribution systems, empty runs occur, which affects the low value of the use of means of transport. Negative effects of transport processes in the city require appropriate solutions enabling their leveling.

Solving transport problems related to the supply of commercial outlets, cleaning the city or transit should include rules enabling coordination of cargo shipment with various means of transport with different power sources, as well as the use of systems and transport organization solutions by consolidating loads in urban areas based on city terminals or logistics centers.

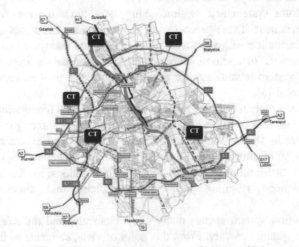

**Fig. 1.** Warsaw agglomeration and city terminals

Activities affecting the organization of transport in distribution systems in urban areas are also among others limitation of delivery time for designated urban areas, restricted traffic areas or the use of a night delivery system. Optimization of goods flows in the city is focused on reducing the time of presence of means of transport in the city, the number and length of routes, as well as better use of loading area of means of transport.

The solution that relieves the city from excessive freight traffic is the construction of logistics centers and cargo handling and consolidation centers - City Terminals. The logistics center makes it possible to eliminate freight transport from the city area, carried out by means of transport with a high load capacity [5]. Another solution is City Terminals, where the coordination of commodity flows in the city takes place in order to combine shipments of specific goods to the center area at one or more centrally located distribution terminals. In addition, city terminals can be used as so-called "Gateway" for the city supply system. To a limited extent they can be cargo consolidation centers for all transport streams flowing through the city. At the same time, in the absence of a logistics center, they can be used directly to service distant transport and be treated as cargo consolidation centers [5, 12, 14, 15]. An example of the Warsaw agglomeration together with the proposal for the deployment of city terminals is shown in Fig. 1.

## 3   Status of the Optimization of Supplies in Distribution Systems

The problem of transport service in urban distribution systems often comes down to the issue of determining optimal routes and scheduling the work of vehicles and crews. The problem of determining optimal transport routes is the subject of studies both in Polish and foreign literature and is called Vehicle routing problem (VRP). This issue is a development of the Travelling Salesman Problem (TSP) [16–19], which also has its own modifications by including additional restrictions and parameters such as vehicle load capacity (CVRP), delivery with time windows (VRPTW), and restrictions on customer demand and collection (VRPB).

Another group of problems is the Vehicle Scheduling Problem (VSP), which is related to the problem of route determination for vehicles (VRP), which in many cases are considered simultaneously. A different group are the problems Crew Sheduling Problem (CSP) or Driver Sheduling Problem (DPS) regarding the issues of scheduling drivers' work at fixed transport tasks [20–23].

In this sense, the Vehicle and Crew Scheduling Problem (VCSP) was formulated and then the benefits of this approach were demonstrated [39]. Problem Fleet Size and Composition Vehicle Routing The problem is closely related to the problem of determining the routes of vehicles. The issue of composition and number of vehicles refers to the selection of vehicles for transport tasks performed taking into account the resources of the enterprise. The criterion is to minimize the total cost of vehicle use as a function of transport tasks performed with limited resources [24–31].

Taking into account significant expenditures on transport means and transport infrastructure as well as operating costs, carriers recognize the role of drivers' scheduling and constructing effective timetables as factors affecting the company's profit and the appropriate level of services. In recent years, these areas have become particularly important due to the growing competition in the transport services market.

An important stage in the modeling process of transport systems is the selection of an appropriate optimization method. The choice of the optimization method is not a simple problem, because there is no one universal method that would be just as effective for all types of optimization tasks of transport systems. The selection of the method of optimization of transport systems should take into account [9, 29–34]:

- type of task parameters - static or dynamic problem,
- number of optimization criteria - a single-criterion or multi-criteria problem,

- type of the optimization problem under consideration - a linear or non-linear problem,
- type of relationship that subordinates task parameters - a deterministic problem, probabilistic or statistic,
- type of decision variables - continuous or discrete variables,
- the existence of restrictions - an issue with or without restrictions.

The algorithms used to solve optimization problems can be divided into classical heuristics and applied heuristics - meta heuristics. The main difference between the above algorithms is based on the fact that the quality of solutions obtained with the use of heuristic meta algorithms is higher than with the use of classic heuristic algorithms, but the time of searching for a solution is longer. Among the classic heuristics there are heuristics of the construction of routes [35], and heuristics of route improvement [31, 36–38]. Meta heuristic algorithms distinguish tabu-search algorithms [39, 40], genetic algorithms [41, 42] simulated annealing algorithms, and ant algorithms [8, 43, 44].

## 4   Mathematical Approach to the Problem

Modeling of the distribution system in urban areas requires its parametrization and formal description, which will define decision variables, constraints and the criterion function, which is a measure of the quality of the solution.

For the purposes of a formal description of the modeling of the urban distribution system, we assume that the transport company has certain means of transport, which form a set of numbers of means of transport, as follows $M = \{m : m = \overline{1,M}\}$. Means of transport are characterized by specific parameters, important from the point of view of transport organization. The parameters are: load capacity $q(m) \in \Re^+, m \in M$ and capacity $g(m) \in \Re^+, m \in M$. In addition, in the system, we will distinguish places of demand for goods (recipients), which form the set $O = \{o : o = \overline{1,O}\}$, where $o = 0$ is urban terminal $(CT)$ and types of goods $R = \{r : r = \overline{1,R}\}$. Consumers' demand was presented in a matrix of the form: $\Xi = [\xi(o,r) \in \Re^+, m \neq 0, o \in O, r \in R]_{O \times R}$. The goods are collected from the municipal terminal and delivered to recipients at specific time intervals. Formulation of the optimization task, also requires the parameterization of the transport network and other characteristics, and selected ones are presented in Table 1. The distribution system model is a static model, often used in planning traffic flows in urban areas. It was assumed that the process of flow of a pipe in the municipal network from the initial to the final place takes place in one long enough time interval. The sensitivity of the model results only from changes in traffic conditions for long time intervals. Due to the adopted assumptions, the model is not sensitive to changes in traffic conditions that occur in the case of a dynamic approach that takes into account the values of the traffic flow characteristics over time.

In order to define decision variables, we assume that the Cartesian product $W \times M \times R$ has a mapping $x$, which performs the elements of this product in the set $\{0, 1\}$, i.e.:

$$x : W \times M \times R \rightarrow \{0, 1\}$$

whereby the size $x((w, w'), m, r) = 1$, section $(w, w')$, is included in the route of the $m$-th mean of transport carrying the $r$-th type of good, otherwise $x((w, w'), m, r) = 0$.

The $x$-mapping can be represented as a matrix:

$$X = \left[ x\left( \left( w, w' \right), m, r \right) \right]_{W \times M \times R}$$

with zero-one variables.

**Table 1.** Parameters of the model.

| Parameter designation | Interpretation of the parameter |
|---|---|
| $\langle \alpha(0, m); \alpha'(0, m) \rangle$ | The time interval in which goods can be collected by the $m$-th mean of transport from $CT, 0 \in O, m \in M$ |
| $\langle \alpha(o, m); \alpha'(o, m) \rangle$ | The time interval in which the $o$-th recipient can accept the cargo, delivered by $m$-th mean of transport, $o \in O, m \in M$ |
| $DTP(m)$ | Daily working time of $m$-th mean of transport, $m \in M$ |
| $\delta1(0, m)$ | Loading time $m$-th mean of transport in $CT, 0 \in O, m \in M$ |
| $\delta2(o, m)$ | Unloading time of $m$-th mean of transport, in $o$-th recipient, $o \in O, m \in M$ |
| $\varepsilon(o, m)$ | The moment of the start of service of $m$-th mean of transport in $o$-th recipient, $o \in O, m \in M$ |
| $\varepsilon'(o, m)$ | The moment of the end of service of $m$-th mean of transport in $o$-th recipient, $o \in O, m \in M$ |
| $dp(w, w')$ | The length of the connection between the vertex $w$ and $w'$ of the transport network |
| (...) | (...) |

For such saved model parameters, should be determined such values of the decision variable $x((w, w'), m, r)$, meeting the limitations:

- each recipient is served by one means,
- one means of transport is sent to each recipient from $CT$,
- routes for means of transport start and end in $CT$,
- excluding cycles that are not part of the set of acceptable solutions,
- customer service in specified time,
- ...
- binary decision variables,

which determine the minimum costs of transport service, recorded as:

$$f(X) = \sum_{w \in W} \sum_{w' \in W} \sum_{m \in M} \sum_{r \in R} dp\left(w, w'\right)\left(x\left(\left(w, w'\right), m, r\right)\right)(k1(m) + k2(m))$$

$$+ \left(\sum_{m \in M} \sum_{w \in W} \sum_{r \in R} \frac{(\varepsilon(o, m) + \tau(o, m) + tp((w, 0), m))(x((w, w'), m, r))}{(k1(m) + k2(m)) - \varepsilon(0, m)}\right) \frac{\psi(m)}{60} \qquad (1)$$

## 5   Case Study

The problem concerns the optimization of the distribution system in the urban agglomeration, which includes the city terminal in Błonie, servicing 75 collection points. The analysis was based on data from one maintenance period, which is one week due to the specifics of the goods. The size of the demand for collection points is expressed in kg, however, due to the specifics of the loads and the parameters of the means of transport, it needs to be expressed in pallet load units.

The transport company realizing deliveries in the distribution system has means of transport with characteristics that enable their use in the urban agglomeration. Two groups of vehicles were analyzed: diesel vehicles and electric vehicles. Other parameters regarding payload and capacity are the same for each group. Calculations for two groups of means of transport were carried out using the proprietary computer program KomWoj, using the heuristic method.

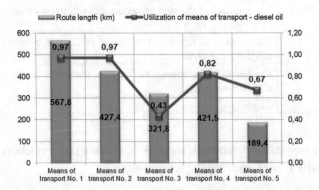

**Fig. 2.** Lengths of routes implemented by means of transport

The total length of routes, both for one and the other group of means of transport, realized in the service of the distribution system for the analyzed settlement period is 1927. 9 km. The smallest one is loaded with means of transport No. 3, whereas the No. 1 is the smallest. There is also a diversification in the rate of utilization of means of transport (entry), which is in the range of 0.41 ÷ 0.97 (Fig. 2).

The length of routes traveled by means of transport in the analyzed distribution system in the urban agglomeration determines transport costs, which also depend on the type of fuel that these means are supplied with. The data on the consumption of particular types of fuel and their unit costs were used for analyzes. Other operating costs were omitted in the analyzes, assuming that these costs are at a similar level for both diesel and electric vehicles. Transport costs for diesel vehicles are higher than for electric vehicles. The analyzes carried out show that the increase of diesel oil consumption by the means of transport by 40%, reduces the transport costs for electric vehicles by 31.5%. With a lower consumption of diesel fuel, at the level of 8 L per 100 km and electricity consumption by the vehicle at 636 Wh/km, the savings amount to 4.1%. The detailed costs for individual means of transport and the reduction of transport costs are shown in Fig. 3.

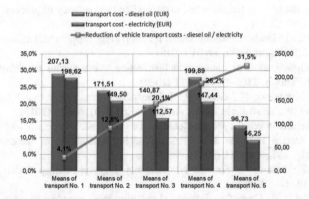

**Fig. 3.** Costs for individual modes of transport and their reduction

# 6   Summary

The problem of distribution of goods in urban agglomerations is an important issue of both practical and research nature. This requires logistic operators to build effective distribution systems dedicated to urban agglomerations. The specificity of urban agglomerations determines the existence of many restrictive conditions for the organization of deliveries, which further affects the search for innovative solutions. In addition to organizational solutions, solutions are also sought for the potential of the transport system, which would allow on the one hand to reduce the emissions of harmful compounds to the environment and, on the other, to minimize transport costs. City distribution systems, due to their complexity and continuous development, require systematic analyzes and research in this area.

# References

1. Pyza, D., Jachimowski, R.: Modelling of parcels' transport system. In: Kersys, R. (ed.) Transport Means, pp. 659–664. Kaunas University of Technology, Kaunas (2015)
2. Pyza, D.: Computer-assisted supply chain distribution processes. In: Grzybowska, K. (ed.) New Insights into Supply Chain, chap. 7, pp. 101–124. Publishing House of Poznan University of Technology, Poznań (2010)

3. Jacyna, M., Merkisz, J.: Proecological approach to modelling traffic organization in national transport system. Arch. Transp. **2**(30), 43–56 (2014)
4. Lewczuk, K., Żak, J., Pyza, D.: Vehicle routing in urban area - environmental and technological determinants. WIT Trans. Built Environ. **130**, 373–384 (2013)
5. Tundys, B.: Logistyka Miejska. Difin, Warszawa (2008)
6. Ambroziak, T., Jachimowski, R., Pyza, D., Szczepański, E.: Analysis of the traffic stream distribution in terms of identification of areas with the highest exhaust pollution. Arch. Transp. **32**(4), 7–16 (2014)
7. Izdebski, M., Jacyna, M.: Use of computer assistance in order to designate the tasks in the municipal services companies. J. Kones **21**(2), 105–112 (2014)
8. Jacyna-Gołda, I., Lewczuk, K.: The method of estimating dependability of supply chain elements on the base of technical and organizational redundancy of process. Maint. Reliab. **19**(3), 382–392 (2017)
9. Jacyna-Gołda, I.: Decision-making model for supporting supply chain efficiency evaluation. Archives of Transport **33**(1), 17–31 (2015)
10. Pyza, D.: Optimization of transport in distribution systems with restrictions on delivery times. Arch. Transp. **21**(3–4), 125–147 (2009)
11. Zieja, M., Smoliński, H., Gołda, P.: Information systems as a tool for supporting the management of aircraft flight safety. Arch. Transp. **36**(4), 67–76 (2015)
12. Pyza, D., Jachimowski, R., Jacyna-Gołda, I., Lewczuk, K.: Performance of equipment and means of internal transport and efficiency of implementation of warehouse processes. Procedia Eng. **187**, 706–711 (2017)
13. Szałek, B.Z.: Miasto w Świetle Nowoczesnej Logistyki. Gospodarka Materiałowa i Logistyka **10**, 14–22 (1995)
14. Pyza, D., Gołda, P.: Cost of environmental pollution in air transport of cargo shipments. Int. J. Energy Sci. **2**(3), 103–109 (2012)
15. Pyza, D.: multicriteria evaluation of designing transportation system within distribution sub-systems. Logist. Transp. **10**(1), 25–34 (2010)
16. Association for the Integration of Warsaw Communication. http://www.siskom.waw.pl
17. Bramel, J., Simchi-Levi, D.: A location based heuristic for general routing problems. Oper. Res. **43**(4), 649–660 (1995)
18. Golden, B., Assad, A., Levy, L., Gheysens, F.: The fleet size and mix vehicle routing problem. Comput. Oper. Res. **11**(1), 49–66 (1984)
19. Jasiński, L.J.: Optymalizacja Dostaw Towarów na Zaopatrzenie Rynku w Warunkach Niepewności. IRWiK, Warszawa (1987)
20. Bektas, T.: The multiple traveling salesman problems and overview of formulations and solution procedures. Omega. Eur. J. Oper. Res. **34**(3), 209–219 (2006)
21. Eglese, R.W., Li, L.: Heuristics for arc routing with a capacity constraint and time deadline. In: Osman, I.H., Kelly, J.P. (eds.) Metaheuristics, Theory and Applications, pp. 633–649. Kluwer, Boston (1996)
22. Fischetti, M., Martello, S., Toth, P.: The fixed job schedule problem with working-time constraints. Eur. J. Oper. Res. **37**(3), 395–403 (1989)
23. Portugal, R., Ramalhinho-Lourenço, H., Paixão, J.P.: Driver scheduling problem modelling. In: Working Paper Series, Social Science Research Network, vol. 991, pp. 123–138 (2007)
24. Clarke, G., Wright, J.W.: Scheduling of vehicles from a central depot to a number of delivery points. Oper. Res. **12**(4), 568–581 (1964)
25. Garcia, B.L., Potvin, J.Y., Rousseau, J.M.: A parallel implementation of the tabu search heuristic for vehicle routing problems with time window constraints. Comput. Oper. Res. **21**(9), 1025–1033 (1994)

26. Gendreau, M., Hertz, A., Laporte, G.: A tabu search heuristic for the vehicle routing problem. Manag. Sci. **40**(10), 1276–1290 (1994)

27. Glover, F.: Future paths for integer programming and links to artificial intelligence. Comput. Oper. Res. **13**(5), 533–549 (1986)

28. Jacyna-Gołda, I., Pyza, D.: Znaczenie Systemów Przewozowych w Łańcuchach Dostaw Przedsiębiorstw Produkcyjnych. Prace Naukowe Politechniki Warszawskiej. Transport **100**, 73–89 (2013)

29. Jacyna-Gołda, I., Izdebski, M., Podviezko, A.: Assessment of efficiency of assignment of vehicles to tasks in supply chains: a case study of a municipal company. Transport **32**(3), 243–251 (2017)

30. Laporte, G., Osman, I.H.: Metaheuristics in combinatorial optimization. Ann. Oper. Res. **63**, 57–75 (1996)

31. Or, I.: Traveling salesman type combinatorial optimization problems and their relation to the logistics of regional blood banking. Northwestern University, Illinois (1967)

32. Ameljańczyk, A.: Optymalizacja Wielokryterialna. Wydawnictwo Wojskowej Akademii Technicznej, Warszawa (1986)

33. Jureczko, M.: Metody Optymalizacji - Przykłady Zadań. Wydawnictwo Pracowni Komputerowej Jacka Skalmierskiego, Gliwice (2009)

34. Pyza, D.: Modelowanie Systemów Przewozowych w Zastosowaniu do Projektowania Obsługi Transportowej Podmiotów Gospodarczych. Oficyna Wydawnicza Politechniki Warszawskiej, Warszawa (2012)

35. Całczyński, A.: Metody Optymalizacyjne w Obsłudze Transportowej Rynku. PWE, Warszawa (1992)

36. Bodin, L., Golden, B., Assad, A., Ball, M.: Routing and scheduling of vehicles and crews. Comput. Oper. Res. **10**(2), 63–211 (1983)

37. Lin, S., Kerninghan, B.W.: An effective heuristic algorithm for travelling salesman problem. Oper. Res. **21**(2), 498–516 (1973)

38. Thompson, P.M., Psaraftis, H.N.: Cyclic transfer algorithms for multi-vehicle routing and scheduling problems. Oper. Res. **41**(5), 935–946 (1993)

39. Freling, R., Wagelmans, A.P.M., Pinto Paixão, J.M.: An overview of models and technics for integrating of vehicle and crew scheduling. In: Wilson, N.H.M. (ed.) Computer Aided Transit Scheduling. LNEMS, vol. 471, pp. 441–460. Springer, Heidelberg (1999)

40. Gheysens, F., Golden, B., Assad, A.: A comparison of techniques for solving the fleet size and mix vehicle routing problem. Oper. - Res. - Spektrum **6**(4), 207–216 (1984)

41. Potvin, J., Bengio, S.: The vehicle routing problem with time windows. Informs J. Comput. **8**(2), 165–172 (1996)

42. Thangiah, S.R.: Vehicle routing with time windows, rusing genetic agorithms. In: Chambers, L. (ed.) Application Handbook of Genetic Algorithms, vol. 2, pp. 254–267. CRC Press, Boca Raton (1995)

43. Russell, R.A.: Hybrid heuristics for the vehicle routing problem with time windows. Transp. Sci. **29**, 156–166 (1995)

44. Jacyna, M.: Some aspects of multicriteria evaluation of traffic flow distribution in a multimodal transport corridor. Arch. Transp. **10**(1–2), 37–52 (1998)

# Using the S-mileSys Tool as a Means to Support Decision Making in Terms of the Fleet Selection for Waste Collection in Urban Areas

Grzegorz Sierpiński[✉]

Faculty of Transport, Silesian University of Technology, Katowice, Poland
grzegorz.sierpinski@polsl.pl

**Abstract.** Linear elements of the transport network may be characterised using numerous factors, including flow capacity, average or permissible running speed, availability to individual means of transport etc. One of the factors that describe the chosen road section is also quality. Quality translates not only into travelling comfort, but also into the negative impact exerted by transport on the environment. Hence the importance of ongoing road quality monitoring. Having up-to-date knowledge in this respect, not only is it possible to plan repairs more effectively, but one can also transfer vehicle streams to routes of higher quality in order to minimise the negative environmental impact of transport (through noise, vibration etc.). This article provides a proposal to use the S-mileSys tool to diagnose technical condition of roads using a fleet of vehicles involved in municipal waste collection. Another stage of the concept implementation is to choose specific vehicles with regard to individual waste collection routes in the transport network in respect of the road quality criterion. What has also been provided in the paper is a case study comprising a small town and a discussion on the potential effect of such efforts as those in question on the given area's transport network. The method proposed in the article may also be applied in another area.

**Keywords:** Waste collection · Fleet assignment · Road quality
S-mileSys

## 1 Introduction

Municipal transport systems always pose a challenge for design engineers, traffic engineers and those in charge of traffic management. The ongoing increase in the number of vehicles translates into major issues encountered while trying to ensure unobstructed traffic flow. One of potential solutions to this problem is to reduce the number of passenger cars in towns by changing the modal split of traffic [1, 2]. Consequently, the number of pedestrians, cyclists and persons using public transport will increase (see also [3–5]). Specific actions aimed at efficiency improvement in transport systems may be conducted on many levels, starting from the micro scale (individual crucial points in the transport network) to the macro scale, when they cover

© Springer Nature Switzerland AG 2019
E. Macioszek and G. Sierpiński (Eds.): Directions of Development of Transport
Networks and Traffic Engineering, LNNS 51, pp. 184–193, 2019.
https://doi.org/10.1007/978-3-319-98615-9_17

entire towns or even a conurbation (examples of micro and macro scale analyses can be found in numerous papers, including [6–13]). However, what also matters besides the growing time loss due to heavy congestion in town centres and urban space occupancy is the environmental aspect [14, 15]. Both the large number of passenger cars and the congestion problem contribute to growing emission of harmful substances and noise. It is possible to reduce such negative environmental impact in various ways, including deployment of low- or zero-emission vehicles.

With regard to freight transport, an attempt to remedy the negative environmental impact would also involve some other actions. Improvement of efficiency (understood not only as cost-effectiveness, but also in the environmental sphere) is also possible through optimisation of goods transport services by combining them into supply chains [16–19]. This problem is particularly evident within the first and the last stage of transport services [20], and mainly concerns town centres.

An aspect often disregarded in planning of routes is road quality. When quality of roads declines, using them also adds to the negative environmental impact of transport due to such factors as noise, vibrations etc. This article comments upon a concept proposed to implement a method enabling technical condition of roads to be monitored and the public service fleet to be selected with reference to individual routes, depending on their quality. The method proposed uses the S-mileSys tool. This tool is one of deliverables of the international project entitled "Smart platform to integrate different freight transport means, manage and foster first and last mile in supply chains" implemented under the ERANET Transport III programme - Sustainable Logistics and Supply Chains [21]. The method has been discussed with reference to a small town case study.

## 2   Characteristics of the S-mileSys Tool

The most fundamental functions performed by the S-mileSys tool is supporting multimodal transport over the first and last mile as well as promotion of eco-friendly means of transport and solutions (Fig. 1).

On the one hand, the tool enables carriers to manage their fleets, and on the other hand, it allows local authorities to oversee transport processes in play within the given area. The tool features numerous implemented optimisation algorithms, including the environment-oriented ones. The built-in route planning criteria take the quality of transport services into consideration.

The tool is composed of multiple modules (six, as shown in Fig. 1). The characterisation below concerns only these modules which can be applied under the method discussed:

- S-mile Freighter Tool is the main S-mileSys component dedicated to transport companies. It is intended for communication and fleet monitoring purposes. From the transport company's perspective, it is the very core of the entire tool, as it provides up-to-date knowledge on the fleet and its location,
- S-mile Transport Planner Tool is a module involved in optimum planning of routes and combining them into chains. The optimisation is based on several criteria, taking time, distance, cost and three environmental sub-criteria into consideration.

With the proposed method in mind, the most important sub-module is Road Condition Tool enabling road quality maps to be created. Road quality is monitored by directly using the given carrier's vehicles. This sub-module contains a mobile application which records such parameters as speed, geographical location and linear acceleration (in three dimensions) as vehicles are on the move,

- S-mile Fleet Management Tool is tool that supports the carrier in decision making by means of algorithms of optimised distribution of goods and transport cost calculation,
- S-mile Visualizer Tool enables visualisation of the information retrieved from Road Condition Tool. It mainly pertains to quality of roads as well as distribution of completed freight transport services over the road network. With such knowledge in disposal, the module can support local authorities in making decisions on traffic organisation within the area subject to analysis, For example, by recommending heavy vehicle traffic restrictions to be introduced in a specific part of town.

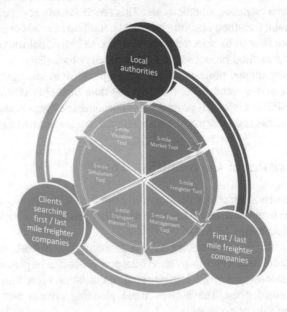

**Fig. 1.** Schematic diagram of the S-mileSys tool

## 3   Method

The S-mileSys tool discussed in the paper may be used to supplement one's knowledge on the quality of roads in the given town without incurring any additional expenses which may be required to conduct advanced and extensive research. What is needed to record the quality of roads is a fleet of vehicles performing daily transport services for the town's purposes. With this goal in mind, the municipal waste management department has been selected. The fleet used to collect and transport waste traverses the town's streets on a daily basis (usually moving along fixed routes). Therefore, this fleet may comprise vehicles recording linear accelerations during transfers.

The following are the preliminary assumptions:

- the waste collection process may be handled in a manner similar to processes connected with what is commonly referred to as the first mile problem,
- the waste collection addresses are the system clients in this case, and the carrier is the waste collection enterprise,
- the vehicle fleet may be defined by such parameters as weight, emission and noise,
- similarly to typical transport processes performed by transport companies, the waste collection process should be optimised in the given town, also by taking environmental criteria into consideration in order to minimise the negative environmental impact of transport.

The implementation process using the S-mileSys tool consists of a number of stages, and it is an iterative process (Fig. 2):

**Fig. 2.** Steps in the implementation of the solution proposed - iterative process

- deploying the Road Condition Tool application in vehicles and data recording. Road Condition Tool is a mobile application running on the Android operating system. It enables monitoring of linear accelerations as the mobile device (i.e. a smartphone) changes its location,
- data analysis. The data thus acquired will make it possible to prepare quality maps with the help of the S-mileSys Visualizer Tool module. And most importantly, these maps can be frequently updated (transfers made on a cyclic basis),
- implementation of repair measures. Making use of the data, local authorities can plan road repairs more effectively. However, they can also minimise the negative environmental impact of transport (which, in this case, is attributed to the waste collection fleet) by applying the road quality criterion. The criteria typically used for optimisation of routes used by the waste collection fleet are those of distance, time or fuel consumption (see also [22–32]). However, taking an additional aspect into account reduces the nuisance experienced by town inhabitants caused by this means of transport.

## 4   Case Study of a Small Town and Potential Effect of Implementation on the Local Transport Network

The case study pertains to a small town of Czeladź (Poland). It extends over the area of ca. 1 640 ha and is inhabited by 33.5 thousand people [33]. The town is a part of the Upper Silesian Conurbation (Fig. 3).

The municipal waste management system covers all the residential real property. There are two independent waste collection schedules functioning in the town: for mixed and for sorted waste. The following are individual characteristics of the municipal waste collection schedule:

- mixed waste:
  - collected twice a month,
  - every consecutive working day, waste is collected from a different part of the town (buildings in specific streets),
- sorted waste:
  - collected once a month,
  - there are 7 pre-defined days in each month when waste is collected from individual areas (buildings in specific streets).

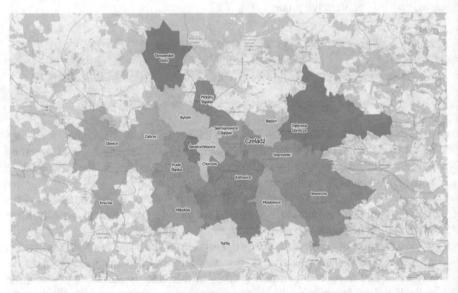

**Fig. 3.** Location of the town of Czeladź in the Upper Silesian Conurbation (Poland) (Source: own research based on [34])

In both cases, the schedule ensures cyclic nature of the road quality measurement, while on the other hand, it displays the need for optimisation of the process in question. This concerns sorted waste collection in particular, since the relevant fleet traverses the given part of the town on each day.

Figures 4 and 5 provide graphical interpretation of the current town areas, as applicable and defined, in a breakdown into consecutive weekdays for both mixed and sorted waste The area covered by the waste collection services performed by the vehicle fleet comprises most streets of the town, which is due to the town's dense housing development.

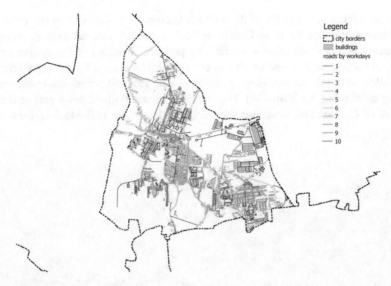

**Fig. 4.** Road network of the area subject to analysis (Czeladź) showing streets covered by the waste collection scheme under the current schedule - mixed waste (Source: own research based on schedule [33])

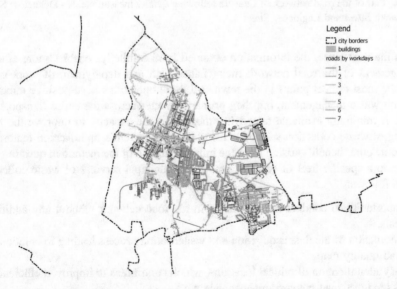

**Fig. 5.** Road network of the area subject to analysis (Czeladź) showing streets covered by the waste collection scheme under the current schedule - sorted waste (Source: own research based on schedule [33])

By furnishing the vehicles of the waste collection fleet addressed in the paper with smartphones featuring the Road Condition Tool application one can identify individual low quality points in the road network. The problems detected in the system may be related to surface defects and cracks, uneven pavement caused by incorrect setting of street inlets or manholes etc. Having collected data from all across the town, one can analyse them using the Visualizer Tool module. Figure 6 illustrates a part of the road network of Czeladź and visualises the quality indicator for individual sections of the network.

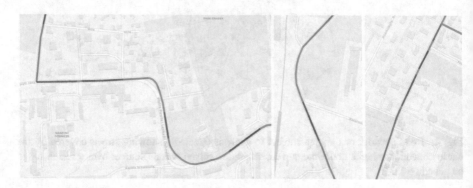

**Fig. 6.** Part of the road network of Czeladź following quality measurements - Dehnelów Street, Katowicka Street and Legionów Street

In the next stage, the information obtained from S-mileSys enables more efficient management of the road network maintenance plan and deployment of work crews over the most critical points in the town. All these operations have positive effect not only on waste management, but they primarily matter across the entire transport network. A minimum available remedying measure is the capacity to improve the functioning of waste collection schedules using the said system's optimisation features.

Some other benefits arising from the implementation of the method in question with regard to a specific fleet of vehicles providing municipal services of waste collection are as follows:

- opportunity to conduct measurements in the road network without any additional costs,
- automation of the data acquisition and visualisation process leading to creation of a road quality map,
- easy identification of critical locations, which contributes to improved efficiency of planning of road network maintenance works,
- possibility to reduce the negative environmental impact of transport, both with reference to the waste collection fleet itself (optimisation of vehicle allocation to individual routes as well as the route selection) and all other aspects of road traffic (by improving the quality of roads via adequate operations of road services).

The disadvantage involved in the application of this system is the incapability of creating speed maps which could dynamically provide information on emergence of congestion in selected sections of the network (which would additionally contribute to better distribution of traffic streams over the road network).

## 5 Conclusions and Further Research

Municipal waste management requires optimisation of the route planning stage as well as of the selection of vehicles for purposes of individual tasks. It is a complex process which often requires application of multiple criteria. A criterion often disregarded is reduction of negative environmental impact.

The method proposed in the article allows for making the most of the fact that waste collection vehicles perform their services on a cyclic basis all over the town area. What the S-mileSys tool enables in this respect is road quality monitoring, which may consequently improve technical condition of road infrastructure. Furthermore, using other modules of the tool in question makes it possible to use the data previously acquired for the sake of optimisation of the delivery process (which, in this case, is waste collection).

Another step in the process of using the tool addressed in the paper is an attempt to integrate it with individual travel planning solutions. Such an approach will facilitate information flow and enable more efficient planning of traffic stream distribution over the transport network.

**Acknowledgements.** The present research has been financed from the means of the National Centre for Research and Development as a part of the international project within the scope of ERA-NET Transport III Programme "Smart platform to integrate different freight transport means, manage and foster first and last mile in supply chains (S-MILE)".

 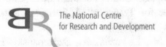 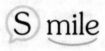

## References

1. Banister, D.: The sustainable mobility paradigm. Transp. Policy **15**, 73–80 (2008)
2. Okraszewska, R., Romanowska, A., Wołek, M., Oskarbski, J., Birr, K., Jamroz, K.: Integration of a multilevel transport system model into sustainable urban mobility planning. Sustainability **10**(2), 1–20 (2018)
3. Turoń, K., Czech, P., Juzek, M.: The concept of walkable city as an alternative form of urban mobility. Sci. J. Silesian Univ. Technol. Ser. Transp. **95**, 223–230 (2017)
4. Nosal, K.: Travel demand management in the context of promoting bike trips, an overview of solutions implemented in Cracow. Transp. Probl. **10**(2), 23–34 (2015)

5. Jacyna, M., Żak, J., Jacyna-Gołda, I., Merkisz, J., Merkisz-Guranowska, A., Pielucha, J.: Selected aspects of the model of proecological transport system. J. KONES, Powertrain Transp. **20**, 193–202 (2013)
6. Macioszek, E.: The comparison of models for follow-up headway at roundabouts. In: Macioszek, E., Sierpiński, G. (eds.) Recent Advances in Traffic Engineering for Transport Networks and Systems. LNNS, vol. 21, pp. 16–26. Springer, Switzerland (2018)
7. Staniek, M.: Moulding of travelling behaviour patterns entailing the condition of road infrastructure. In: Macioszek, E., Sierpiński, G. (eds.) Contemporary Challenges of Transport Systems and Traffic Engineering. LNNS, vol. 2, pp. 181–191. Springer, Cham (2017)
8. Piecha, J., Staniek, M.: The context-sensitive grammar for vehicle movement description. In: Bolc, L., Tadeusiewicz, R., Chmielewski, L.J., Wojciechowski, K. (eds.) Computer Vision and Graphics, Part II. LNCS, vol. 6375, pp. 193–202. Springer, Berlin (2010)
9. Staniek, M.: Stereo vision method application to road inspection. Baltic J. Road Bridge Eng. **12**(1), 38–47 (2017)
10. Macioszek, E., Lach, D.: Analysis of the results of general traffic measurements in the West Pomeranian Voivodeship from 2005 to 2015. Sci. J. Silesian Univ. Technol. Ser. Transp. **97**, 93–104 (2017)
11. Stanley, J.: Land use/transport integration: starting at the right place. Res. Transp. Econ. **48**, 381–388 (2014)
12. Galińska, B.: Multiple criteria evaluation of global transportation systems - analysis of case study. In: Sierpiński, G. (ed.) Advanced Solutions of Transport Systems for Growing Mobility. AISC, vol. 631, pp. 155–171. Springer, Cham (2018)
13. Celiński, I.: Transport network parameterisation using the GTAlg tool. In: Macioszek, E., Sierpiński, G. (eds.) Contemporary Challenges of Transport Systems and Traffic Engineering. LNNS, vol. 2, pp. 111–123. Springer, Cham (2017)
14. European Commission: White Paper: Roadmap to a Single European Transport Area - Towards a Competitive and Resource Efficient Transport System. European Commission, Luxembourg (2011)
15. European Commission: Communication from the Commission to the European Parliament, the Council, the European Economic and Social Committee and the Committee of the Regions: Clean Power for Transport: A European Alternative Fuels Strategy. European Commission, Brussels (2013)
16. Wasiak, M., Jacyna, M., Lewczuk, K., Szczepański, E.: The method for evaluation of efficiency of the concept of centrally managed distribution in cities. Transport **32**(4), 348–357 (2017)
17. Kijewska, K., Małecki, K., Iwan, S.: Analysis of data needs and having for the integrated urban freight transport management system. In: Mikulski, J. (ed.) Challenge of Transport Telematics. CCIS, vol. 640, pp. 135–148. Springer, Cham (2016)
18. Hoang Son, L., Louati, A.: Modeling municipal solid waste collection: a generalized vehicle routing model with multiple transfer stations, gather sites and inhomogeneous vehicles in time windows. Waste Manag. **52**, 34–49 (2016)
19. Buhrkal, K., Larsen, A., Ropke, S.: The waste collection vehicle routing problem with time windows in a city logistics context. Procedia – Soc. Behav. Sci. **39**, 241–254 (2012)
20. Macioszek, E.: First and last mile delivery - problems and issues. In: Sierpiński, G. (ed.) Advanced Solutions of Transport Systems for Growing Mobility. AISC, vol. 631, pp. 147–154. Springer, Cham (2018)
21. Smart Platform to Integrate Different Freight Transport Means, Manage and Foster First and Last Mile in Supply Chains. ERA-NET Transport III. Sustainable Logistics and Supply Chains. Project Proposal (2015)

22. Ferreira, F., Avelino, C., Bentes, I., Matos, C., Teixeira, C.A.: Assessment strategies for municipal selective waste collection schemes. Waste Manag. **59**, 3–13 (2017)
23. Ghiani, G., Laganà, D., Manni, E., Musmanno, R., Vigo, D.: Operations research in solid waste management: a survey of strategic and tactical issues. Comput. Oper. Res. **44**, 22–32 (2014)
24. Faccio, M., Persona, A., Zanin, G.: Waste collection multi objective model with real time traceability data. Waste Manag. **31**, 2391–2405 (2011)
25. Das, S., Bhattacharyya, B.K.: Optimization of municipal solid waste collection and transportation routes. Waste Manag. **43**, 9–18 (2015)
26. Hoang Son, L.: Optimizing municipal solid waste collection using chaotic particle swarm optimization in GIS based environments: a case study at Danang city, Vietnam. Expert Syst. Appl. **41**, 8062–8074 (2014)
27. Malakahmad, A., Bakri, P.M., Mokhtar, M.R.M., Khalil, N.: Solid waste collection routes optimization via GIS techniques in Ipoh city, Malaysia. Procedia Eng. **77**, 20–27 (2014)
28. Buenrostro-Delgado, O., Ortega-Rodriguez, J.M., Clemitshaw, K.C., González-Razo, C., Hernández-Paniagua, I.Y.: Use of genetic algorithms to improve the solid waste collection service in an urban area. Waste Manag. **41**, 20–27 (2015)
29. Laureri, F., Minciardi, R., Robba, M.: An algorithm for the optimal collection of wet waste. Waste Manag. **48**, 56–63 (2016)
30. Lee, C.K.M., Yeung, C.L., Xiong, Z.R., Chung, S.H.: A mathematical model for municipal solid waste management - a case study in Hong Kong. Waste Manag. **58**, 430–441 (2016)
31. Jaunich, M.K., Levis, J.W., De Carolis, J.F., Gaston, E.V., Barlaz, M.A., Bartelt-Hunt, S.L., Jones, E.G., Hauser, L., Jaikumar, R.: Characterization of municipal solid waste collection operations. Resour. Conserv. Recycl. **114**, 92–102 (2016)
32. Nguyen-Trong, K., Nguyen-Thi-Ngoc, A., Nguyen-Ngoc, D., Dinh-Thi-Hai, V.: Optimization of municipal solid waste transportation by integrating GIS analysis, equation-based, and agent-based model. Waste Manag. **59**, 14–22 (2017)
33. Czeladź City. http://www.czeladz.pl
34. Open Street Map. http://www.openstreetmap.org/

# Author Index

© Springer Nature Switzerland AG 2019
E. Macioszek and G. Sierpiński (Eds.): Directions of Development of Transport
Networks and Traffic Engineering, LNNS 51, p. 195, 2019.
https://doi.org/10.1007/978-3-319-98615-9

Printed in the United States
By Bookmasters